Materials for Electrochemical Energy Conversion and Storage

Related titles published by The American Ceramic Society:

Boing-Boing the Bionic Cat and the Jewel Thief
By Larry L. Hench
©2001, ISBN 1-57498-129-3

Boing-Boing the Bionic Cat
By Larry L. Hench
©2000, ISBN 1-57498-109-9

The Magic of Ceramics
By David W. Richerson
©2000, ISBN 1-57498-050-5

Perovskite Oxides for Electronic, Energy Conversion, and Energy Efficiency Applications (Ceramic Transactions, Volume 104)
Edited by Winnie Wong-Ng, Terry Holesinger, Gil Riley, and Ruyan Guo
©2000, ISBN 1-57498-091-2

Processing and Characterization of Electrochemical Materials and Devices (Ceramic Transactions, Volume 109)
Edited by Prashant N. Kumta, Arumugam Manthiram, S.K. Sundaram, and Yet-Ming Chiang
©2000, ISBN 1-57498-096-3

Electrochemistry of Glass and Ceramics (Ceramic Transactions, Volume 92)
Edited by S.K. Sundaram, Dennis F. Bickford, and E.J. Hornyak Jr.
©1999, ISBN 1-57498-062-9

Ceramic Innovations in the 20th Century
Edited by John B. Wachtman Jr.
©1999, ISBN 1-57498-093-9

For information on ordering titles published by The American Ceramic Society, or to request a publications catalog, please contact our Customer Service Department at 614-794-5890 (phone), 614-794-5892 (fax),<customersrvc@acers.org> (e-mail), or write to Customer Service Department, 735 Ceramic Place, Westerville, OH 43081, USA.

Visit our on-line book catalog at <www.ceramics.org>.

Ceramic
Transactions
Volume 127

Materials for Electrochemical Energy Conversion and Storage

Papers from the Electrochemical Materials, Processes, and
Devices symposium at the 102nd Annual Meeting of The
American Ceramic Society, held April 29–May 3, 2000, in St.
Louis, Missouri, and the Materials for Electrochemical Energy
Conversion and Storage symposium at the 103rd Annual
Meeting of The American Ceramic Society, held April 22–25,
2001, in Indianapolis, Indiana, USA.

Edited by

Arumugam Manthiram
University of Texas at Austin

Prashant N. Kumta
Carnegie Mellon University

S.K. Sundaram
Pacific Northwest National Laboratory

Gerbrand Ceder
Massachusetts Institute of Technology

Published by
The American Ceramic Society
735 Ceramic Place
Westerville, Ohio 43081
www.ceramics.org

Papers from the Electrochemical Materials, Processes, and Devices symposium at the 102nd Annual Meeting of The American Ceramic Society, held April 29–May 3, 2000, in St. Louis, Missouri, and the Materials for Electrochemical Energy Conversion and Storage symposium at the 103rd Annual Meeting of The American Ceramic Society, held April 22–25, 2001, in Indianapolis, Indiana, USA.

Cover photo: "SEM micrograph of the cross-section of a fuel cell fracture surface showing the dense electrolyte between the porous anode and cathode layers," is courtesy of S. Huss, R. Doshi, J. Guan, G. Lear, K. Montgomery, N. Minh, and E. Ong, and appears as figure 4 in the paper "Materials and Microstructures for Improved Solid Oxide Fuel Cells," which begins on page 109.

Library of Congress Cataloging-in-Publication Data
A CIP record for this book is available from the Library of Congress.

For information on ordering titles published by The American Ceramic Society, or to request a publications catalog, please call 614-794-5890.

ISBN 978-1-57498-135-3

Contents

Fuel Cells

Lithium-Ion Batteries

Preface

Environmental concerns and the continuing global depletion of fossil fuels have created enormous worldwide interest in alternative energy technologies. Electrochemical power sources such as batteries, fuel cells, and supercapacitors have become appealing in this regard as they offer clean energy.

Batteries are the major power sources for portable electronic devices. The miniaturization and exponential growth of popular portable electronic devices such as cellular phones and laptop computers have created an ever-increasing demand for compact lightweight power sources. Lithium-ion batteries have become appealing in this regard as they offer higher energy density compared to other rechargeable systems. The higher energy density also makes them attractive for electric vehicles. Commercial lithium-ion cells are currently made with the layered lithium cobalt oxide cathode (positive electrode) and carbon anode (negative electrode). Unfortunately, structural and chemical instability of cobalt oxide limits the practical capacity to 50% of its theoretical capacity. Additionally, cobalt is expensive and toxic. The challenge is to develop inexpensive and environmentally benign cathode hosts with high electrochemical capacity and good safety characteristics. With respect to the anode, the currently used graphite tends to exhibit irreversible capacity losses. There is immense interest to develop alternative cathode and anode materials.

Fuel cells are attractive for both stationary and portable power and for electric vehicles. Unlike battery technology, fuel-cell technology has not quite matured and is confronted with materials issues and high cost. For example, the yttria-based solid oxide fuel cell requires a high operating temperature of around 1000°C, which limits the choice of electrode and interconnect materials. The solid oxide fuel cells currently use lanthanum strontium manganate cathode, which, due to its poor oxide-ion conductivity, leads to significant over-potential and performance losses. Development of structurally and chemically stable mixed conductors with acceptable oxide–ion conduction and compatible thermal expansion characteristics as cathode is needed. Alternatively, successful development of intermediate temperature (700°C) electrolytes such as the lathanum strontium gallate and compatible electrodes could have a significant impact. Additionally, cost-effective manufacturing is critical for the success of fuel-cell technology.

In addition to being used as electrodes in solid oxide fuel cells, mixed conductors exhibiting both oxide–ion and electronic conduction also find applications as oxygen gas separation membranes. However, good structural and chemical stabilities under the operating conditions of around 900°C and reducing environment with high oxygen permeation flux are essential for such applications. While the perovskite oxides are the most widely investigated materials, they often tend to exhibit structural instability due to oxygen losses. In this regard, some perovskite-related intergrowth oxides have become appealing recently.

To bring the ceramics community up to date on these technologies, the American Ceramic Society has been hosting symposia on this topic since 1995. This volume consists of five papers that were presented at the 102nd Annual Meeting of the American Ceramic Society, St. Louis, Missouri, April 30–May 3, 2000 and 20 papers presented at the 103rd Annual Meeting of the American Ceramic Society, Indianapolis, Indiana, April 22–25, 2001. A number of leading experts in materials science and engineering, solid state chemistry and physics, electrochemical science and technology from academia, industry and national laboratories presented their work at these symposia. The presentations covered development of new materials and a fundamental understanding of the structure-property-performance relationships and the associated electrochemical and solid state phenomena. The symposia were sponsored by the Electronics, Basic Science, and Nuclear and Environmental divisions of the American Ceramic Society. They were also co-sponsored by the Electrochemical Society. A total of 73 and 56 papers were presented, respectively, at the 2000 and 2001 symposia including several invited talks. Among them, 25 peer-reviewed papers are included in this volume. After first presenting the five papers from the 2000 symposium, the 20 papers from the 2001 symposium are grouped under three subtopics: gas separation membranes, fuel cells, and lithium-ion batteries.

The editors acknowledge the support of several members of the Electronics, Basic Science, and Nuclear and Environmental divisions of the American Ceramic Society. The editors also thank all the authors, session chairs, manuscript reviewers, and the society staff who made the symposia and the proceedings volume a success. It is the sincere hope of the editors that the readers will appreciate and benefit from this collection of articles in the area of electrochemical energy conversion and storage, some of which were from world-renowned experts in materials and electrochemical science and engineering.

Arumugam Manthiram
Prashant N. Kumta
S.K. Sundaram
Gerbrand Ceder

Papers from 2000 Meeting

OXYGEN PERMEATION PROPERTIES OF THE INTERGROWTH OXIDE $Sr_{3-x}La_xFe_{2-y}Co_yO_{7-\delta}$

F. Prado, T. Armstrong, and A. Manthiram
Texas Materials Institute, ETC 9.104
The University of Texas at Austin
Austin, TX 78712

A. Caneiro
Centro Atómico Bariloche, CNEA
8400- S. C. de Bariloche
Argentina

ABSTRACT

The structural stability and oxygen permeation properties of the perovskite-related intergrowth oxides $Sr_{3-x}La_xFe_{2-y}Co_yO_{7-\delta}$ with x = 0.0 and 0.3 and y = 0.0 and 0.6 have been studied at high temperature (800 ≤ T ≤ 900°C) and in the oxygen partial pressure (pO$_2$) range between air and pure N$_2$. Samples with x = 0 show instability on exposure to ambient air conditions while samples with x = 0.3 are stable. High temperature X-ray diffraction data reveal no phase transitions for these compounds at 900 °C in pure nitrogen. The volume expansions of the $Sr_{3-x}La_xFe_{2-y}Co_yO_{7-\delta}$ oxides due to changes in oxygen content at 900 °C are lower than those for the perovskite phase $SrCo_{0.8}Fe_{0.2}O_{3-\delta}$. Oxygen permeation flux values for the $Sr_{3-x}La_xFe_{2-y}Co_yO_{7-\delta}$ compounds are approximately one order of magnitude lower than those for the perovskite phase $SrCo_{0.8}Fe_{0.2}O_{3-\delta}$. The oxygen permeation flux increases with increasing Co content, but decreases with the substitution of La for Sr.

INTRODUCTION

In recent years, mixed conductors exhibiting oxide-ion and electronic conduction have received much attention due to their possible uses in many applications. Examples include their use as electrodes in solid oxide fuel cells, catalysis, and oxygen separation membranes [1]. As these materials exhibit both electronic and oxide-ion conductivity, oxygen permeation across a dense ceramic membrane is possible in the presence of a pO$_2$ gradient without the use of electrodes and external circuitry required for a traditional ceramic oxygen pump. Such membranes can, for example, be used for the direct conversion of methane to synthetic gas (CO+H$_2$) using air as the oxygen source. However, several problems need to be addressed since these applications require materials with high oxygen permeability as well as good structural and chemical stability at high temperatures and in reducing atmospheres.

High oxygen permeability values have been reported for perovskite phases with the general formula $La_{1-x}A_xCo_{1-y}B_yO_{3-\delta}$ (A = alkaline or rare earth and B = Fe, Ni and Cu) [2,3]. These compounds are the most widely investigated mixed conductors as they have high concentration of oxygen vacancies ($\delta \approx 0.2 - 0.5$), which facilitate oxide-ion diffusion, and exhibit high electronic conductivity ($\sim 10^2-10^3$ S cm^{-1}). However, as oxygen vacancy concentrations increase, electrostatic interactions also increase which escalates the likelihood of a structural transformation to an ordered structure. Furthermore, the ordered structure could transform to a disordered structure on heating to higher temperatures. These structural changes along with chemical instability of the perovskite phases in reducing atmospheres have been suggested as the cause for low mechanical integrity of perovskite membranes [4,5].

One possible solution to the structural instability of perovskite phases is to develop perovskite-related intergrowth oxides as oxygen separation membranes. For example, strontium iron oxides with iron in the 3+/4+ oxidation state form a Ruddlesden-Popper (RP) series having the general formula $Sr_{n+1}Fe_nO_{3n+1}$. The crystal structure of this series can be visualized as layers stacked along the z axis

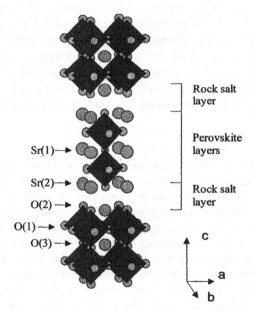

Figure 1: Crystal structure of $Sr_3Fe_2O_{7-\delta}$.

with n SrFeO$_3$ perovskite layers alternating with rock-salt SrO layers. The $n = 2$ member of the series, the Sr$_3$Fe$_2$O$_{7-\delta}$ phase (Figure 1), exhibits oxygen nonstoichiometry ranging between $0 \leq \delta \leq 1$ without changes in the crystal structure [6]. This phase stability makes Sr$_3$Fe$_2$O$_{7-\delta}$ a possible candidate for use as an oxygen separation ceramic membrane. However, only limited information is available on cationic substitutions and no information is available in the literature with regard to the oxygen permeation properties of the Sr$_3$Fe$_2$O$_{7-\delta}$ phase. The objective of this work is to synthesize Sr$_{3-x}$La$_x$Fe$_{2-y}$Co$_y$O$_{7-\delta}$ oxides and investigate the structural stability and oxygen permeation properties of this new class of compounds. The results are compared with those of the perovskite phase SrFe$_{0.2}$Co$_{0.8}$O$_{3-\delta}$.

EXPERIMENTAL

Powdered samples of Sr$_3$Fe$_{2-y}$Co$_y$O$_{7-\delta}$ with y = 0.0 and 0.6 and SrFe$_{0.2}$Co$_{0.8}$O$_{3-\delta}$ were prepared by a solid state reaction of SrCO$_3$, Fe$_2$O$_3$ and Co$_3$O$_4$. Stoichiometric amounts of the raw materials were ground and calcined at 900 °C for 12 h in air. The powder mixture was ground again and pressed into discs that are 2 mm thick and 24 mm in diameter and fired at 1300 °C for 24 h. The final sintering temperature for the SrFe$_{0.2}$Co$_{0.8}$O$_{3-\delta}$ perovskite phase was 1150 °C. Lanthanum doped Sr$_3$Fe$_2$O$_{7-\delta}$ samples were not single phase when prepared by solid state reaction and they contained significant amounts of the RP members with n = 1 and n = 3. However, lanthanum doped samples containing only a small fraction of perovskite phase (< 3 %) could be synthesized by a sol-gel method [7]. The starting materials for the sol-gel method were La$_2$O$_3$, SrCO$_3$, Fe(CH$_3$COO)$_2$ and Co(CH$_3$COO)$_2$. The gel was decomposed at 400 °C for 10 min in air and the resulting powder was pressed into a disc and fired at 1400 °C for 12 h. The densities of the sintered discs produced by both methods were found to be > 90 % of the theoretical densities.

X-ray powder diffraction data at room temperature were collected on a Phillips APD 3520 diffractometer using Cu Kα radiation. For high temperature X-ray powder diffraction, samples were spread on a resistively heated platinum ribbon mounted in an Anton Paar HTK-10 camera coupled to Phillips PW1700 diffractometer. Rietveld refinements were carried out with the DBWS-9411 program [8]. Thermogravimetric analysis was performed with a Perkin-Elmer Series 7 Thermal Analyzer with a heating rate of 1°C/min.

Oxygen permeation measurements as a function of time at constant temperature and He flow rate, and as a function of temperature at constant pO$_2$ gradient were performed in the temperature range $800 \leq T \leq 900$ °C using a permeation setup that utilizes flowing He as a carrier gas and a Hewlett Packard

5880A gas chromatograph. The details of the oxygen permeation setup have been previously reported [7].

RESULTS AND DISCUSSION

The crystal structures of $Sr_3Fe_2O_{7-\delta}$ and $Sr_3Fe_{1.4}Co_{0.6}O_{7-\delta}$ synthesized by solid state reaction and $Sr_{2.7}La_{0.3}Fe_{1.4}Co_{0.6}O_{7-\delta}$ synthesized by sol-gel method were refined at room temperature on the basis of the tetragonal space group I4/mmm using the Rietveld method. The experimental data, calculated profile, and difference between the two are shown in Fig. 2. Samples without La were single phase while the La doped sample showed the presence of a trace amount of perovskite phase (< 3 %). The phase analysis results and lattice parameters obtained at room temperature and 900 °C are given in Table I for the various compositions.

In situ X-ray diffraction data at 900 °C of $Sr_3Fe_2O_{7-\delta}$, $Sr_{2.7}La_{0.3}Fe_{1.4}Co_{0.6}O_{7-\delta}$ and $SrFe_{0.2}Co_{0.8}O_{3-\delta}$ reveal no structural transitions for any of the compounds on going from air to pure N_2. A volume expansion of 0.12% was obtained from the lattice parameter values for $Sr_3Fe_2O_{7-\delta}$ at 900 °C when the pO_2 changes from air

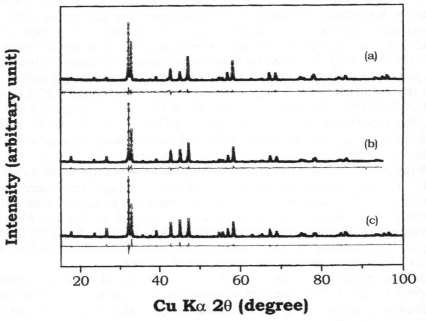

Figure 2: X-ray powder diffraction data, calculated profile, and their difference for (a) $Sr_3Fe_2O_{7-\delta}$, (b) $Sr_3Fe_{1.4}Co_{0.6}O_{7-\delta}$, (c) $Sr_{2.7}La_{0.3}Fe_{1.4}Co_{0.6}O_{7-\delta}$.

Table I: Phase analysis and lattice parameters at 25 °C and 900 °C of $Sr_{3-x}La_xFe_{2-y}Co_yO_{7-\delta}$ ($0.0 \leq x \leq 0.3$ and $0.0 \leq y \leq 0.6$) and $SrFe_{0.2}Co_{0.8}O_{3-\delta}$.

Composition	Temperature /atmosphere	Space Group	Lattice Parameters and Unit Cell Volume			Impurity Phases [a]
			a(Å)	c(Å)	V(Å³)	
$Sr_3Fe_2O_{7-\delta}$	25°C/ air	I4/mmm	3.868(1)	20.158(2)	301.7(2)	-
	900 °C/ air	I4/mmm	3.929(1)	20.413(2)	315.1(2)	-
	900 °C/ N₂	I4/mmm	3.942(1)	20.303(2)	315.5(2)	-
$Sr_3Fe_{1.4}Co_{0.6}O_{7-\delta}$	25°C/ air	I4/mmm	3.854(1)	20.148(2)	299.3(2)	-
$Sr_{2.7}La_{0.3}Fe_{1.4}Co_{0.6}O_{7-\delta}$	25 °C/ air	I4/mmm	3.853(1)	20.138(2)	299.0(2)	P < 3 %
	900 °C/ air	I4/mmm	3.928(1)	20.447(2)	315.5(2)	P < 3 %
	900 °C/ N₂	I4/mmm	3.941(1)	20.445(2)	317.5(2)	P < 3 %
$SrFe_{0.2}Co_{0.8}O_{3-\delta}$	900 °C/ air	Pm-3m	3.946(1)	-	61.44(5)	CoO
	900 °C/ N₂	Pm-3m	3.962(1)	-	62.19(5)	CoO

[a] P: Perovskite phase

to N₂ while the change was 0.65 % for $Sr_{2.7}La_{0.3}Fe_{1.4}Co_{0.6}O_{7-\delta}$. As the pO₂ decreases, the *a* parameter increases in both the samples, while the *c* parameter decreases in $Sr_3Fe_2O_{7-\delta}$ and remains almost constant in $Sr_{2.7}La_{0.3}Fe_{1.4}Co_{0.6}O_{7-\delta}$. In the case of the perovskite phase $SrFe_{0.2}Co_{0.8}O_{3-\delta}$, a volume expansion of 1.24 % was found under similar conditions and this value is half the value reported by Pei et al. [4]. Furthermore, the high temperature X-ray diffraction data reveal that the crystal structure of $SrFe_{0.2}Co_{0.8}O_{3-\delta}$ at 900 °C remains cubic on changing the pO₂ from air to pure N₂, which is in agreement with Qiu et al. [9]. However, thermogravimetric analysis of the perovskite phase $SrFe_{0.2}Co_{0.8}O_{3-\delta}$ performed in pure N₂ shows a pronounced oxygen loss between 400 °C and 500 °C and a plateau between 500 °C and 850 °C due to the stabilization of the brownmillerite phase $Sr_2Co_{1.6}Fe_{0.4}O_5$ as confirmed by X-ray diffraction. Thus, the $SrCo_{0.8}Fe_{0.2}O_{3-\delta}$ perovskite phase transforms to an orthorhombic brownmillerite phase at 500 °C which further transforms to a cubic phase at T ≈ 850 °C. On the other hand, the $Sr_{3-x}La_xFe_{2-y}Co_yO_{7-\delta}$ compounds do not exhibit any phase transformation with temperature in either air or pure N₂.

Figure 3 shows the oxygen permeation flux at 900 °C as a function of time for $Sr_3Fe_2O_{7-\delta}$, $Sr_3Fe_{1.4}Co_{0.6}O_{7-\delta}$ and $Sr_{2.7}La_{0.3}Fe_{1.4}Co_{0.6}O_{7-\delta}$. Measurements were performed with 1.5 mm thick samples with one side of the sample exposed to air (pO₂' = 0.209 atm) and the other side (pO₂") exposed to a constant He flow rate of 10 ml/min. All three samples exhibited stable oxygen permeation flux values over a period of ~ 4 days. For a given Co content, partial substitution of La^{3+} for Sr^{2+} decreases the oxygen permeation flux. However, La^{3+} substitution was beneficial to suppress the structural and chemical instability or reactivity at room

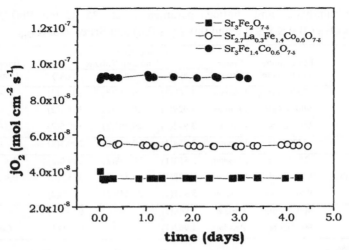

Figure 3. Variation of oxygen permeation flux with time for $Sr_3Fe_2O_{7-\delta}$ (■), $Sr_{2.7}La_{0.3}Fe_{1.4}Co_{0.6}O_{7-\delta}$ (O) and $Sr_3Fe_{1.4}Co_{0.6}O_{7-\delta}$ (●). Measurements were conducted at 900 °C with 1.5 mm thick samples and at a constant He flow rate of 10 ml/min.

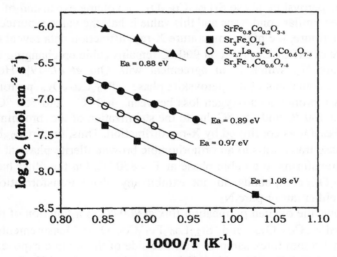

Figure 4. Arrhenius plot of the oxygen permeation flux for $Sr_3Fe_2O_{7-\delta}$ (■), $Sr_3Fe_{1.4}Co_{0.6}O_{7-\delta}$ (●), and $Sr_{2.7}La_{0.3}Fe_{1.4}Co_{0.6}O_{7-\delta}$ (O) at $\log(pO_2'/pO_2'') = 2.2$ and the perovskite phase $SrFe_{0.2}Co_{0.8}O_{3-\delta}$ (▲) at $\log(pO_2'/pO_2'') = 1.4$.

temperature. Samples without La were unstable in air at room temperature and degrade with time due to a slow reaction with H_2O or CO_2. On the other hand, the substitution of Co for Fe in the $Sr_{3-x}La_xFe_{2-y}Co_yO_{7-\delta}$ phase increased the oxygen flux. This increase in oxygen permeation with Co doping is a consequence of both higher electronic conductivity and higher oxygen vacancy concentration.

Figure 4 shows the Arrhenius plots of the oxygen permeation flux for $Sr_3Fe_2O_{7-\delta}$, $Sr_3Fe_{1.4}Co_{0.6}O_{7-\delta}$, $Sr_{2.7}La_{0.3}Fe_{1.4}Co_{0.6}O_{7-\delta}$ and $SrFe_{0.2}Co_{0.8}O_{3-\delta}$ with a constant pO_2 gradient across the membrane. The pO_2 difference for the intergrowth compounds was $\log(pO_2'/pO_2'') = 2.2$, while the pO_2 difference for the perovskite phase was $\log(pO_2'/pO_2'') = 1.4$. The difference in the oxygen permeation flux values between the intergrowth phases and the perovskite phase are approximately one order of magnitude. An apparent activation energy associated with the permeation process was calculated from the slope of the line in the Arrhenius plot in the temperature range of $800 \leq T \leq 900$ °C. The activation energy values of $0.89 \leq E_a \leq 1.08$ eV obtained for the $Sr_{3-x}La_xFe_{2-y}Co_yO_{7-\delta}$ intergrowth oxides are similar to those reported for the $La_{1-x}Sr_xFe_{1-y}Co_yO_{3-\delta}$ perovskite compounds [1,10]. Thus, the lower oxygen permeation flux values exhibited by the $Sr_{3-x}La_xFe_{2-y}Co_yO_{7-\delta}$ phases compared to the perovskite phase $SrFe_{0.2}Co_{0.8}O_{3-\delta}$ might be related to a lower oxygen vacancy concentration and the two-dimensional character of the intergrowth phase.

CONCLUSIONS

The perovskite-related intergrowth oxides $Sr_{3-x}La_xFe_{2-y}Co_yO_{7-\delta}$ were synthesized by solid state reaction and a sol-gel technique. Samples without lanthanum substitution are single phase while those doped with lanthanum show the presence of a trace amount of a perovskite phase (< 3 %). However, samples doped with lanthanum are stable when exposed to ambient air conditions while samples without lanthanum showed signs of decomposition. Under the conditions of oxygen permeation at $800 \leq T \leq 900$ °C, the $Sr_{3-x}La_xFe_{2-y}Co_yO_{7-\delta}$ oxides are chemically stable without undergoing any phase transformation in the pO_2 range from air to pure N_2. Furthermore, the $Sr_3Fe_2O_{7-\delta}$ compounds with La and Co doping experience a volume expansion less than that for the perovskite phase $SrCo_{0.8}Fe_{0.2}O_{3-\delta}$ when the pO_2 varies from air to pure N_2 at 900 °C. The oxygen permeation flux values for the $Sr_{3-x}La_xFe_{2-y}Co_yO_{7-\delta}$ intergrowth oxides are approximately one order of magnitude lower than those for the $SrFe_{0.2}Co_{0.8}O_{3-\delta}$ perovskite phase. The oxygen permeation flux for the $Sr_{3-x}La_xFe_{2-y}Co_yO_{7-\delta}$ membranes are constant with time, decrease with increasing La content x and increase with increasing Co content y.

ACKNOWLEDGEMENT

Acknowledgment is made to the Welch Foundation (Grant No. F-1254) and the donors of Petroleum Research Fund administered by the American Chemical Society (Grant No. ACS-PRF#32410-AC5) for the support of this research. One of the authors (F. P.) thanks CONICET, Argentina, for a postdoctoral fellowship.

REFERENCES

[1] H. J. M. Bouwmeester and A. J. Burggraaf, "Dense Ceramic Membranes for Oxygen Separation"; pp 481-553 in *The CRC Handbook of Solid State Electrochemistry*, Edited by P. J. Gellings and H. J. M. Bouwmeester. CRC Press, 1997.

[2] Y. Teraoka, H. M. Zhang, S. Furukawa, and N. Yamazoe, "Oxygen Permeation Through Perovskite-Type Oxides," *Chemistry Letters*, 1743-1746 (1985).

[3] Y. Teraoka, T. Nobunaga and N. Yamazoe, "Effect of Cation Substitution on the Oxygen Semipermeability of Perovskite Oxides," *Chemistry Letters*, 503-506 (1988).

[4] S. Pei, M. S. Kleefisch, T. P. Kobylinski, K. Faber, C. A. Udovich, V. Zhang-McCoy, B. Dabrowski, U. Balachandran, R.L. Mieville, and R. B. Poeppel, "Failure Mechanisms of Ceramics Membrane Reactors in Partial Oxidation of Methane to Synthesis Gas," *Catalysis Letters*, **30**, 210-212 (1995).

[5] U. Balachandran, J. T. Dusek, S. M. Sweeney, R. B. Poeppel, R. L. Mieville, P. S. Maiya, M. S. Kleefisch, S. Pei, T. P. Kobylinski, C. A. Udovich and A. C. Bose, "Methane to Syngas via Ceramics Membranes," *American Ceramic Society Bulletin*, **74**, 71-75 (1995).

[6] S. E. Dann, M. T. Weller, and D. B. Curie, "Structure and Oxygen Stoichiometry in $Sr_3F_2O_{7-\delta}$," *Journal of Solid State Chemistry*, **97**, 179-185 (1992).

[7] T. Armstrong, F. Prado, Y. Xia, and A. Manthiram, "Role of Perovskite Phase on the Oxygen Permeation Properties of the $Sr_4Fe_{6-x}Co_xO_{13+\delta}$ System," *Journal of the Electrochemical Society*, **147**, 435-438 (2000).

[8] R.A. Young, A. Sakthivel, T. S. Moss, C. O. Paiva Santos, "DBWS-9411 program for Rietveld refinement," *Journal of Applied Crystallography*, **28**, 366-367(1995).

[9] L. Qiu, T. H. Lee, L.-M. Liu, Y.L. Yang, and A. J. Jacobson, "Oxygen Permeation Studies of $SrCo_{0.8}Fe_{0.2}O_{3-\delta}$," *Solid State Ionics*, **76** , 321 (1995).

[10] S. Li, W. Jin, P. Huang, N. Xu, J. Shi, Y. S. Lin, M. Z. -C. Hu and E. A. Payzant, "Comparison of Oxygen Permeability and Stability of Perovskite Type $La_{0.2}A_{0.8}Co_{0.2}Fe_{0.8}O_{3-\delta}$ (A = Sr, Ba, Ca) Membranes," *Industrial Engineering Chemistry Research*, **38**, 2963-2972 (1999).

STRUCTURAL STABILITY OF LITHIUM EXTRACTED $Li_{1-x}Ni_{1-y}Co_yO_2$

R. V. Chebiam, F. Prado, and A. Manthiram
Texas Materials Institute, ETC 9.104
The University of Texas at Austin
Austin, TX 78712

ABSTRACT

The structural stability of $Li_{1-x}Ni_{1-y}Co_yO_2$ ($0 \leq y \leq 0.3$) cathodes has been investigated by chemically extracting lithium with an oxidizing agent from the layered $LiNi_{1-y}Co_yO_2$ and heating the samples at moderate temperatures (50-150 °C). The delithiated samples experience a decrease in the c/a ratio at temperatures as low as 50 °C due to a migration of the $Ni^{3+/4+}$ ions from the Ni layer to the Li layer. The change in the c/a ratio increases with increasing temperature and degree of delithiation, but decreases with increasing cobalt content y. Thus the $Li_{1-x}CoO_2$ cathodes are structurally stable while the $Li_{1-x}Ni_{1-y}Co_yO_2$ cathodes suffer from a structural instability at moderate temperatures.

INTRODUCTION

The higher energy density of lithium-ion cells compared to other systems has made them attractive for portable electronic devices [1]. Commercial lithium-ion cells use $LiCoO_2$ as the cathode and carbon as the anode. $LiCoO_2$ offers a capacity of approximately 140 mAh/g with excellent electrochemical performance. Also, it can be readily synthesized as a stoichiometric material by conventional methods [2, 3]. However, the high cost and toxicity of cobalt necessitates the development of alternate materials. $LiNiO_2$ is a possible candidate as Ni is less expensive and less toxic than Co, but $LiNiO_2$ suffers from Jahn-Teller distortion [4, 5] and irreversible phase transitions during the charge-discharge cycling [6-8]. Also, $LiNiO_2$ is more difficult to synthesize. More recently, these difficulties have been suppressed by a partial substitution of Co for Ni and compositions such as $LiNi_{0.85}Co_{0.15}O_2$ offer a much higher capacity (180 mAh/g) than $LiCoO_2$ [9-11]. However, the structural and chemical stability of nickel-based cathodes during long term cycling particularly at moderate temperatures (50-100 °C) and high rates remain to be established. We present in this paper a systematic investigation of the structural stability of chemically delithiated $Li_{1-x}Ni_{1-y}Co_yO_2$ by monitoring the lattice parameters on heating at moderate temperatures. The data are compared with that of $LiCoO_2$ cathodes.

EXPERIMENTAL

$LiNi_{1-y}Co_yO_2$ ($0 \leq y \leq 0.3$) oxides were synthesized by a sol-gel procedure [12]. Required amounts of lithium carbonate, nickel acetate and cobalt acetate were dissolved in acetic acid and refluxed. After about an hour, small amounts of hydrogen peroxide and water were added and the mixture was refluxed for another hour. The solution was then dried slowly on a hot plate to yield a semi-transparent gel, which was initially decomposed at 400 °C for 10 min in air and then fired at 750 - 800 °C for 24 hours in an oxygen atmosphere. $LiCoO_2$ was synthesized by solid-state reaction between Li_2CO_3 and Co_3O_4 in air at 900 °C for 24 hours. Chemical extraction of lithium from $LiNi_{1-y}Co_yO_2$ and $LiCoO_2$ was performed in aqueous and non-aqueous media respectively. In the case of aqueous medium, the sample was stirred with an aqueous solution of the oxidizer, sodium perdisulfate, for 3 days. During this, the following reaction occurs:

$$2\ LiNi_{1-y}Co_yO_2 + x\ Na_2S_2O_8 \rightarrow 2\ Li_{1-x}Ni_{1-y}Co_yO_2 + x\ Na_2SO_4 + x\ Li_2SO_4$$

The product was filtered, washed with water first and then acetone, and allowed to dry. In the case of non-aqueous medium, the sample was stirred with an acetonitrile solution of the oxidizer, NO_2PF_6, for 2 days [13]. The product was washed with acetonitrile and dried under vacuum.

The samples were characterized by X-ray powder diffraction. The lattice parameters and cation distribution were refined by Rietveld method using the DBWS-9006 PC program [14]. Lithium contents were determined by Atomic absorption spectroscopy.

RESULTS AND DISCUSSIONS

X-ray diffraction and atomic absorption spectroscopic data reveal that $Li_{1-x}Ni_{1-y}Co_yO_2$ samples could be obtained for $0 \leq (1-x) \leq 0.35$ by extracting lithium in aqueous medium; for $(1-x) < 0.35$, $\gamma\text{-}Ni_{1-y}Co_yOOH$ is formed. Extraction of 0.65 lithium corresponds to a capacity of approximately 180 mAh/g. Extraction of lithium from $LiCoO_2$ in aqueous medium with sodium perdisulfate resulted in the formation of a second phase $Na_{0.6}CoO_2$ for lithium contents $(1-x) < 0.4$ in $Li_{1-x}CoO_2$. Therefore, the $Li_{1-x}CoO_2$ samples were obtained by extracting lithium in acetonitrile medium with NO_2PF_6. X-ray diffraction shows that the lithium extracted samples maintain the rhombohedral layer structure in both the systems.

Figure 1 shows the variation of a and c lattice parameters of $Li_{0.37}Ni_{0.9}Co_{0.1}O_2$ with heating time at various temperatures (50-150 °C). The a parameter increases and the c parameter and c/a ratio decrease as the samples are

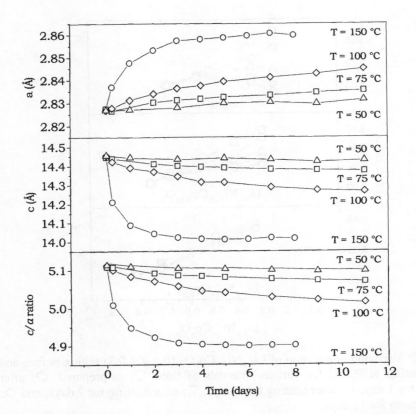

Figure 1: Variations of a and c parameters and c/a ratio of $Li_{0.37}Ni_{0.9}Co_{0.1}O_2$ with heating time at various temperatures.

heated. The rate of change of lattice parameters increases as the heating temperature increases and the c/a ratio approaches the ideal cubic-close-packing value of 4.9 on heating at 150 °C for a few days; *i.e.* the sample transforms from rhombohedral to cubic symmetry on heating at 150 °C. This is also evident in the X-ray diffraction patterns by a merger of the (108) and (110) reflections and (006) and (012) reflections.

Figure 2 shows the variation of c/a ratio with the degree of delithiation for $Li_{1-x}Ni_{1-y}Co_yO_2$ ($0 \leq y \leq 0.3$) after heating at 70 °C in air for various periods of time. The data show that the change in c/a ratio becomes more and more pronounced as the degree of delithiation increases or lithium content $(1-x)$ decreases. Also, for a given lithium content, the change in c/a ratio decreases with increasing Co content y.

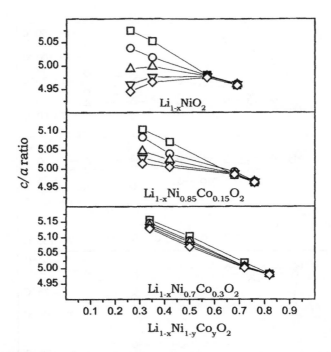

Figure 2: Variation of *c/a* ratio of $Li_{1-x}Ni_{1-y}Co_yO_2$ ($0 \leq y \leq 0.3$) with x before and after heating at 70 °C for various intervals of time. $\square$: as-prepared, $\bigcirc$: after heating for 1 day, $\triangle$: after heating for 3 days, $\triangledown$: after heating for 7 days, and $\Diamond$: after heating for 11 days.

Figure 3 compares the variations of *c/a* ratio with heating time at a moderate temperature of 70 °C for various cobalt contents y in $Li_{1-x}Ni_{1-y}Co_yO_2$ having nearly the same degree of delithiation (lithium content (1-x) = 0.33 ± 0.02). The results show that the tendency to transform towards the cubic symmetry decreases with increasing cobalt content. A substitution of as low as 30 % Co for Ni suppresses significantly the tendency to transform from the rhombohedral layer structure to the cubic structure. More importantly, there is hardly any decrease in the *c/a* ratio with heating time for the cobalt oxide $Li_{0.35}CoO_2$ under similar conditions illustrating the structural stability of cobalt oxides compared to the nickel oxides. The data clearly show that the delithiated nickel oxides experience a structural instability at moderate temperatures, which may have consequences in the long-term cyclability of these cathodes.

Materials for Electrochemical Energy Conversion and Storage

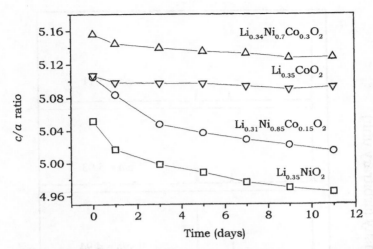

Figure 3: Variations of c/a ratio with heating time at 70 °C for various cobalt contents in $Li_{0.33\pm0.02}Ni_{1-y}Co_yO_2$.

With an objective to identify the origin of the decrease in the c/a ratio, we carried out a Rietveld analysis of the $Li_{0.37}Ni_{0.9}Co_{0.1}O_2$ composition before and after heating at 150 °C for various periods of time. The refinements were carried out with $R\bar{3}m$ space group and the refinement results are given in Figure 4 and Table I. The as-prepared sample does not have any Ni in the Li layer (3a sites). With an increase in the heating time, the number of Ni atoms in the Li layer increases indicating a migration of the Ni atoms from the Ni layer (3b sites) to the Li layer (3a sites). The migration of Ni from the 3b octahedral sites to the 3a octahedral sites can occur through the available tetrahedral sites. The structural stability of the delithiated $Li_{1-x}CoO_2$ compared to $Li_{1-x}Ni_{1-y}Co_yO_2$ reveals that while $Ni^{3+/4+}$ are able to migrate through the tetrahedral sites, such a migration of the $Co^{3+/4+}$ ions is energetically unfavorable.

With an objective to see whether the cubic sample obtained after heating at 150 °C for 5 days has the spinel structure, the X-ray data was also refined with the spinel structure (Table I). Such a refinement with Li^+ in the 8a tetrahedral sites and $(Ni,Co)^{3+/4+}$ in the 16d octahedral site yielded a lattice parameter of 8.082(1) Å and a R_{wp} of 11.7 %. However, the quality of the refinement was found to improve when the $(Ni,Co)^{3+/4+}$ ions were allowed to occupy partially the empty 16c octahedral sites of the spinel lattice. A lower R_{wp} value of 10.5 % was obtained when 15 % of the $(Ni,Co)^{3+/4+}$ ions were allowed to occupy the 16c sites

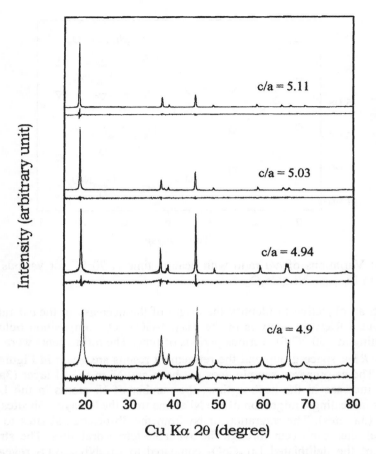

Figure 4: Rietveld refinement of $Li_{0.37}Ni_{0.9}Co_{0.1}O_2$ after heating at 150 °C for various periods of time.

and the remaining 85 % to occupy the 16d sites. Transformation from a rhombohedral layer structure to an ideal cubic spinel structure requires a migration of 25 % of the transitional metal ions from the 3b octahedral sites to the 3a octahedral sites and a displacement of the Li^+ ions into the neighboring 8a tetrahedral sites. The presence of 15 % of the $(Ni,Co)^{3+/4+}$ ions in the 16c octahedral sites in the present case indicates that the migration of the $(Ni,Co)^{3+/4+}$ ions from the Ni layer to the Li layer is incomplete at the lower heating temperature of 150 °C.

Table I. Cation distributions of $Li_{0.37}Ni_{0.9}Co_{0.1}O_2$ before and after heating at 150 °C for various times

Heating time (days)	Cation distribution	
	Layer structure ($R\bar{3}m$)	Spinel structure ($Fd\bar{3}m$)
0	$[Li_{0.37}]_{3a}[(Ni, Co)]_{3b}O_2$	-
< 1	$[Li_{0.37}(Ni, Co)_{0.025}]_{3a}[(Ni, Co)_{0.975}]_{3b}O_2$	-
1	$[Li_{0.37}(Ni, Co)_{0.09}]_{3a}[(Ni, Co)_{0.91}]_{3b}O_2$	-
5	-	$(Li_{0.74})_{8a}[(Ni, Co)_{0.29}]_{16c}[(Ni, Co)_{1.71}]_{16d}O_4$

CONCLUSIONS

The structural stability of delithiated $Li_{1-x}Ni_{1-y}Co_yO_2$ has been assessed and compared with that of $Li_{1-x}CoO_2$ by heating at moderate temperatures. The $Li_{1-x}Ni_{1-y}Co_yO_2$ cathodes experience a structural instability at temperatures as low as 50 °C due to a migration of the transition metal ions from the Ni layer to the Li layer while the $Li_{1-x}CoO_2$ cathodes are quite stable under similar conditions. Similar results were found with analogous experiments carried out with charged coin cells indicating that electrochemically delithiated samples also behave similar to the chemically delithiated samples. The structural instability of the nickel-based cathodes may pose difficulties in the long-term cyclability of lithium-ion cells particularly at higher temperatures or under high rates.

ACKNOWLEDGMENTS

This work was supported by Center for Space Power at the Texas A&M University (a NASA Commercial Space Center), Texas Advanced Technology Program Grant 003658-0488-1999 and the Welch Foundation Grant F-1254.

REFERENCES

[1]B. Scrosati, "Challenge of Portable Power," *Nature*, **373**, 557-558 (1995).

[2]K. Mizushima, P. C. Jones, P. J. Wiseman and J. B. Goodenough, "Li_xCoO_2 (0 < x < 1): A New Cathode Material for Batteries of High Energy Density," *Materials Research Bulletin*, **15**, 783- 799 (1980).

[4]I. Nakai and T. Nakagome, "In Situ Transmission X-Ray Absorption Fine Structure Analysis of the Li Deintercalation Process in $Li(Ni_{0.5}Co_{0.5})O_2$," *Electrochemical and Solid State Letters,* **1**, 259 (1998).

[5]I. Nakai, K. Takahashi, Y. Shiraishi, T. Nakagome, and F. Nishikawa, "Study of the Jahn-Teller Distortion in $LiNiO_2$, a Cathode Material in a Rechargeable Lithium Battery, by *in Situ* X-Ray Absorption Fine Structure Analysis," *Journal of Solid State Chemistry,* **140**, 145 (1998).

[6]W. Li, J. N. Reimers, and J. R. Dahn, "In-Situ X-ray Diffraction and Electrochemical Studies of $Li_{1-x}NiO_2$," *Solid State Ionics*, **67**, 123, (1993).

[7]T. Ohzuku, A. Ueda, and M. Nagayama, "Electrochemistry and Structural Chemistry of $LiNiO_2$ ($R\overline{3}m$) for 4 Volt Secondary Lithium Cells," *Journal of the Electrochemical Society*,**140**, 1862 (1993).

[8]C. Delmas, J. P. Peres, A. Rougier, A. Demourgues, F. Weill, A. Chadwick, M. Broussely, F. Perton, Ph. Biensan, and P. Willmann, "On the Behavior of the Li_xNiO_2 system: An Electrochemical and Structural Overview," *Journal of Power Sources*, **68**, 120 (1997).

[9]W. Li and J. C. Currie, "Morphology Effects on the Electrochemical Performance of $LiNi_{1-x}Co_xO_2$," *Journal of the Electrochemical Society,* **144**, 2773 (1997).

[10]J. Cho, G. B. Kim, and H. S. Lim, "Effect of Preparation Methods of $LiNi_{1-x}Co_xO_2$ Cathode Materials on Their Chemical Structure and Electrode Performance," *Journal of the Electrochemical Society,* **146**, 3571 (1999).

[11]J. Cho, H. S. Jung, Y. C. Park, G. B. Kim, and H. S. Lim, "Electrochemical Properties and Thermal Stability of $Li_aNi_{1-x}Co_xO_2$ Cathode Materials," *Journal of the Electrochemical Society,* **147**, 15 (2000).

[12]T. Armstrong, F. Prado, Y. Xia, and A. Manthiram, "Role of Perovskite Phase on the Oxygen Permeation Properties of the $Sr_4Fe_{6-x}Co_xO_{13+\delta}$ System," *Journal of the Electrochemical Society,* **147**, 435 (2000).

[13]A. R. Wizansky, P. E. Rauch and F. J. DiSalvo, "Powerful Oxidizing Agents for the Oxidative Deintercalation of Lithium from Transition metal Oxides," *Journal of Solid State Chemistry,* **81**, 203-207 (1989).

[14]R. A. Young, A. Shakthivel, T. S Moss, and C. O. Paiva Santos, "DBWS-9411 program for Rietveld refinement," *Journal of Applied Crystallography,* **28**, 366 (1995).

Reaction Kinetics and Mechanism of Formation of LiNiO$_2$ From Particulate Sol-Gel (PSG) Derived Precursors

C. C. Chang, J. Y. Kim, Z. G. Yang and P. N. Kumta
Department of Materials Science and Engineering
Carnegie Mellon University, Pittsburgh, PA 15213

ABSTRACT

The reaction kinetics and mechanism of formation of LiNiO$_2$ from particulate sol-gel (PSG) derived precursors have been investigated using a combination of X-ray diffraction (XRD) and thermo gravimetric analyses (TGA) techniques. It has been confirmed that Li$_2$CO$_3$ and NiO are the intermediate products formed after decomposing the xerogel. Subsequent heat treatment enables further reaction between Li$_2$CO$_3$ and NiO to form LiNiO$_2$. A single peak observed in the differential weight loss analysis suggests a direct reaction between Li$_2$CO$_3$ and NiO with simultaneous evolution of CO$_2$. The activation energy associated with this direct reaction is also estimated as 98.5 kJ/mole using the non-isothermal Kissinger's method. A comparison of the extent of completion of the reactions conducted on both PSG and solid state derived precursors indicate a faster reaction kinetics of formation of LiNiO$_2$ using the PSG process.

INTRODUCTION

Lithiated transition metal oxides LiMO$_2$ (M= Mn, Ni, Co, Ni$_x$Co$_{1-x}$) are technically important cathode materials for lithium-ion battery applications because they possess high energy density and capacity [1-11]. These materials are 2-D intercalation compounds which have a layered structure with Li$^+$ cations inserted in between the MO$_2^-$ (M= transition metal cations) slabs. In the case of LiNiO$_2$, our earlier studies have shown that the precursors prepared from the particulate sol-gel (PSG) process are potentially effective for synthesizing LiNiO$_2$ at moderate temperature and short time [12, 13]. However, the reaction mechanisms and kinetics involved in the formation of LiNiO$_2$ from the PSG derived xerogels are not yet clear. The present paper thus focuses on the identification of the reaction mechanisms responsible for the formation of LiNiO$_2$ using the PSG process. The paper also reports the evaluation of activation energies associated with the reactions and finally demonstrates the faster kinetics of formation of LiNiO$_2$ for the PSG process in comparison to the solid state methods.

EXPERIMENTAL

The xerogel utilized for synthesizing LiNiO$_2$ and conducting the kinetic study was prepared using lithium hydroxide monohydrate (Aldrich, 99 %) and proper stoichiometric amount of nickel acetate tetrahydrate (Aldrich, 98 %), as starting materials. Generally, 0.1 mole of lithium and 0.1 mole of nickel were first dissolved in de-ionized (DI) water separately to obtain clear solutions. Mixing of the individual solutions resulted in pale green colored suspensions (180 ml volume). Ethyl alcohol (90 ml) was then added to the solutions to facilitate removal of the liquid products. The pale green turbid solutions were stirred for 15-20 minutes immediately prior to drying.

A rotary evaporator (Buchi) was used for the subsequent drying process employing an initial pressure of 500 mbar at 120°C for 3 hours followed by a reduced pressure of 100 mbar at 140°C for 1 hour to dry the solutions completely. Heat treatments were conducted in air in the temperature range of 300°C ~ 800°C at increments of 100°C each, using alumina boats as sample carriers. The dwell time for all the as-prepared powders was set at 5 hours for each of the temperatures in the range of 300°C ~ 800°C.

The heat-treated powders were ground and sieved through 325 mesh (aperture width *ca.* 30 μm in size) for X-ray diffraction (XRD, Rigaku θ/θ diffractometer) characterization. A Cu X-ray tube was consistently used as the radiation source. A 0.05° step size at 35 kV and 20 mA with a duration time of 2 seconds were the general parameters used for the XRD analysis. Thermo-Gravimetric/Differential thermal analysis was also conducted on the xerogel using the simultaneous TG/DTA (TA Instruments, model 2960). The heat treatment schedule was varied depending on the experimental objectives and will be specified in the following text when necessary.

RESULTS AND DISCUSSION

1. *Formation of lithium carbonate and nickel oxide after decomposition:*

Fig. 1 shows the result of TGA analysis conducted on the xerogel sample. The corresponding XRD study of the xerogel samples heat treated from 300 to 800°C showing the evolution of the phases is indicated in Fig. 2. From Fig. 1, a significant weight loss (43.04%) before 300°C followed by a further moderate weight loss (7.9%) before 800°C are observed. Correspondingly, the presence of NiO and Li_2CO_3, and the formation of $LiNiO_2$ are also observed for the samples heat treated between 300 and 700°C as shown by the results of the phase evolution study. The first step weight loss observed in the TGA results can be correlated to the decomposition of the xerogel based on the information obtained from FTIR and chemical analysis of the xerogel. The decomposition reaction can be written as follows:

$$xLiOH + (1-x)/2\ Ni(OH)_2 + \qquad 300°C$$
$$(1-x)LiOAc + (1+x)/2\ Ni(OAc)_2 \ \text{---------}> 1/2\ Li_2CO_3 + NiO + gases \qquad (1)$$
$$44.2\% \text{ weight loss calculated, } 43.04\% \text{ observed}$$

However, if the second step weight loss is attributed to the reaction between NiO and Li_2CO_3, the expected weight loss should be 7%:

$$300 \text{ to } 800°C$$
$$NiO + 1/2\ Li_2CO_3 + 1/4\ O_{2(g)} \ \text{---------}> LiNiO_2 + 1/2\ CO_{2(g)} \uparrow \qquad (2)$$
$$7.0\ \% \text{ weight loss calculated, } 7.9\% \text{ observed}$$

Thus the calculated weight loss and the observed weight loss are in good agreement (7% vs. 7.9%). Furthermore the difference between the calculated total weight loss and the observed total weight loss (51.2% vs. 50.94%) is even smaller. One can therefore conclude that $LiNiO_2$ is formed at elevated temperatures due to the reaction between NiO and Li_2CO_3 formed after the decomposition of the xerogel.

2. *Does lithium nickel oxide form via a direct single step or a multi-step reaction?*

According to the conclusion drawn earlier, the overall reaction between lithium carbonate and nickel oxide to form lithium nickel oxide can always be written as:

$$2NiO + Li_2CO_3 + 1/2\ O_{2(g)} \ \text{---------}> 2LiNiO_2 + CO_{2(g)} \uparrow \qquad (3)$$

However, it is also possible for the above reaction to proceed via two steps:

$$Li_2CO_3 \ \text{---------}> Li_2O + CO_{2(g)} \uparrow \qquad (4)$$

Materials for Electrochemical Energy Conversion and Storage

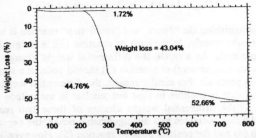

Fig. 1 shows the result of the TGA analysis of the xerogel sample. A significant weight loss (43.04%) prior to 300°C followed by a further moderate weight loss (7.9%) before 800°C are observed.

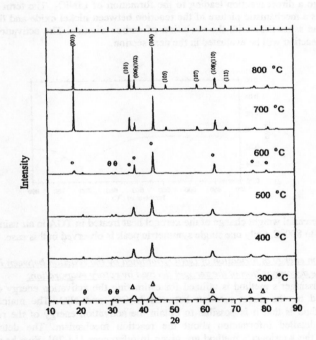

Fig. 2. The XRD results showing the phase evolution of LiNiO$_2$ during heat treatment of the PSG-derived xerogels. -Δ- indicates the NiO phase. -θ- and -o- represent Li$_2$CO$_3$ and LiNiO$_2$ phases, respectively.

$$1/2\ O_{2(g)} + Li_2O + 2\ NiO \longrightarrow 2\ LiNiO_2 \qquad (5)$$

The major factor distinguishing the "direct" and "multi-step" reaction is to infer whether reactions (4) and (5) occur simultaneously. In the case of reaction (3), a weight loss should be observed while the reaction proceeds. As a result, the differential weight change (dM/dT, 'M' represents mass and 'T' represents temperature) of reaction (3) should possess a single peak that is negative in sign as the reaction proceeds. For similar reason, the differential weight change of reaction (5) should exhibit a peak that is positive in sign but smaller in magnitude than reaction (4) as the reaction progresses. If the differential weight change of these two reactions between lithium carbonate and nickel oxide possesses two distinguishable peaks, the negative peak can always be attributed to reaction (4) and the positive peak to reaction (5). However, if the differential weight change between lithium carbonate and nickel oxide exhibits only one peak, then it is clearly indicative of the simultaneous occurrence of reaction (4) and reaction (5). This is because the superposition of a negative and a positive peak can result in a single peak only if they are located at the same position, which implies the simultaneous occurrence of the two reactions (4) and (5).

The differential weight change of the xerogel heat treated in TGA in air using a heating rate of 10°C/min up to 800°C is shown in Fig. 3. A single symmetric peak is observed in this case which clearly implies that lithium carbonate and nickel oxide obtained by decomposition of the xerogel undergo a direct reaction leading to the formation of LiNiO₂. The term 'direct' reaction used here gives a mechanistic picture of the reaction between nickel oxide and lithium carbonate that react at the same time with the evolution of carbon dioxide. The activation energy of the overall direct reaction will be evaluated in the next section.

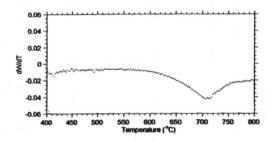

Fig. 3. The differential weight change of the xerogel heat treated in TGA in air using a heating rate of 10°C/min up to 800°C. Only one single symmetric peak is observed in this case.

3. *The activation energy of formation of LiNiO₂ obtained by the reaction between Li₂CO₃ and NiO generated by the decomposition of the xerogel derived by rotary evaporation:*

The Kissinger's method is utilized for estimating the activation energy of formation of LiNiO₂ obtained by the direct reaction between Li₂CO₃ and NiO. The major advantage of Kissinger's method is that it is possible to obtain the activation energy of the reaction without requiring any detailed information about the reaction mechanism. The detailed principles associated with the Kissinger's method are given in references [14-20]. Five heating rates were chosen for the TGA analysis. The 5 different heating rates chosen for heat treating the samples from room temperature to 900°C in air are 5, 10, 20, 30 and 40°C/min respectively. The results of the TGA analysis are shown in Fig. 4.

Materials for Electrochemical Energy Conversion and Storage

Since the reaction of interest occurs in the temperature range corresponding to the completion of the initial decomposition reaction (complete formation of Li_2CO_3 and NiO), the differential weight loss versus temperature of these five samples shown in Fig. 5(a) arises at 500°C and above. From Fig. 5(a), the reaction maximum temperature (T_M) can be determined for each sample that was heat treated at each of the different heating rates. The reaction maximum temperature (T_M) for the samples that were heat treated using the heating rates of 5, 10, 20, 30 and 40°C/min are thus determined to be 690°C, 709 °C, 738 °C, 758 °C and 771 °C respectively. It should be noticed that the peaks seen after 800°C can be attributed to the decomposition of $LiNiO_2$ [21-23]. The activation energy for the formation of $LiNiO_2$ from the reaction between Li_2CO_3 and NiO can now be calculated by obtaining the slope of the plot of $\ln[(dT/dt)/T_M^2]$ versus $1/T_M$. The plot $\ln[(dT/dt)/T_M^2]$ versus $1/T_M$ is shown in Fig. 5(b). Linear regression of these points, yields the slope of $-11866 = -Ea/R$. The activation energy for formation of $LiNiO_2$ via the reaction of Li_2CO_3 and NiO obtained by decomposition of the xerogel in a rotary evaporator therefore determined to be 98.65 kJ/mole.

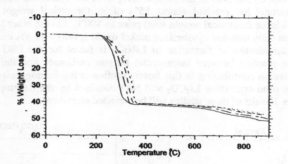

Fig. 4. The results of the TGA analysis for the 5 different heating rates chosen for heat treating the samples up to 900°C in air. The heating rates are 5, 10, 20, 30 and 40°C/min respectively.

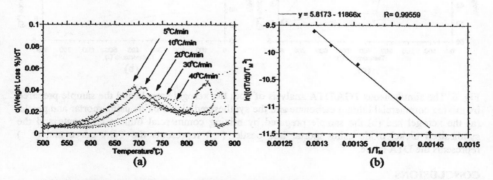

Fig. 5. (a)The differential weight loss versus temperature of samples heat treated in air for the 5 different heating rates. (b) The plot of $\ln[(dT/dt)/T_M^2]$ versus $1/T_M$ gives the slope,'-Ea/R from which the activation energy (Ea) can be calculated.

4. A comparative study of reaction kinetics: the formation of LiNiO₂ via the PSG derived xerogel precursor and the solid state reaction of a mixture of commercial Li₂CO₃ and NiO.

4. A comparative study of reaction kinetics: the formation of LiNiO₂ via the PSG derived xerogel precursor and the solid state reaction of a mixture of commercial Li₂CO₃ and NiO.

A comparative study of the kinetics of $LiNiO_2$ formation via the reaction between Li_2CO_3 and NiO obtained from both the decomposition of the xerogel and the conventional solid state process is conducted again using TGA analysis. Fig. 6(a) and (b) show the results of the simultaneous TGA/DTA analysis conducted on both the xerogel and the sample prepared by mixing commercial lithium carbonate and the synthesized nickel oxide using a mortar and pestle. The NiO used here was synthesized by decomposing commercially available $Ni(OH)_2$ at 800°C in air for 2 h. Results shown in Fig. 6(a) and (b) were obtained using a heating rate of 10°C/min to 800°C, followed by a dwell time of 1 minute and a cooling rate of 10°C/min back to room temperature. From the DTA analysis of the xerogel shown in Fig. 6(a), it can be seen that no distinct endotherm or exotherm is observed representing melting or solidification of lithium carbonate during the heating as well as the cooling process. In contrast, the DTA analysis of the sample consisting of commercial lithium carbonate and synthesized nickel oxide shows both melting and solidification of lithium carbonate (see Fig. 6(b)). This result suggests a faster reaction between Li_2CO_3 and NiO in the xerogel sample. A more detailed weight analysis has also shown that the xerogel has reached almost 85% of its theoretical weight loss (according to equation (2), 7.0% is the theoretical weight loss) prior to 800°C. On the other hand, the reaction of commercial lithium carbonate and synthesized nickel oxide is only ~41.3% complete for the same condition. Thus, the kinetics of formation of $LiNiO_2$ is faster for the PSG derived xerogel in comparison to the reaction between commercial lithium carbonate and the synthesized nickel oxide. The main reason contributing to this faster reaction is the homogeneity and the nanoscale distribution of the nano crystalline Li_2CO_3 and NiO obtained by the decomposition of the PSG derived precursors. Results of these studies will be reported elsewhere.

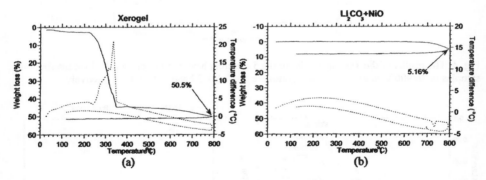

Fig. 6. The simultaneous TGA/DTA analysis of both the xerogel sample and the sample prepared by mixing commercial lithium carbonate and the synthesized nickel oxide using mortar and pestle. (a) the xerogel and (b) the sample prepared by mixing commercial lithium carbonate and the synthesized nickel oxide using mortar and pestle. (—) represents the TGA curve and (·····) represents the DTA curve.

CONCLUSIONS

The reaction mechanism and kinetics of formation of $LiNiO_2$ utilizing the PSG derived xerogel has been successfully conducted. The xerogel precursor first undergoes a decomposition reaction to form the intermediate products Li_2CO_3 and NiO. Subsequent heating renders the formation of $LiNiO_2$ via the direct reaction between Li_2CO_3 and NiO. The direct reaction between Li_2CO_3 and NiO with simultaneous evolution of CO_2 is unequivocally shown by analyzing the differential weight loss plot obtained from the TGA analysis of the xerogel. The activation energy associated with the reaction has also been successfully determined to be 98.5 kJ/mole using the non-isothermal Kissinger's method. The faster reaction kinetics of the PSG derived xerogel compared to the solid state reaction of the same starting components demonstrates the significant benefits of using the PSG derived xerogel as the precursor for synthesizing $LiNiO_2$ at moderate temperature and in a short time.

ACKNOWLEDGMENT

The authors gratefully acknowledge the support of Changs Ascending, Taiwan for this research. NSF (Grant CTS-9700343) and the technical assistance of Eveready Battery Company is also acknowledged.

REFERENCES

[1] M. Broussely, F. Perton, P. Biensan, J. M. Bodet, J. Labat, A. Lecerf, C. Delmas, A. Rougier, and J. P. Peres, *J. Power Sources*, **54**, 109 (1995).

[2] T. Ohzuku, A. Ueda, M. Nagayama, Y. Iwakoshi and H. Komori, *Electrochim. Acta*, **38**, 1159 (1993).

[3] B. Banov, J. Bourilkov, and M. Mladenov, *J. Power Sources*, **54**, 268 (1995).

[4] D. Caurant, N. Baffier, B. Garcia, and J. P. Pereira-Ramos, *Solid State Ionics*, **91**, 45 (1996).

[5] T. Ohzuku, A. Ueda, and M. Nagayama, *J. Electrochem. Soc.*, **140**, 1862 (1993).

[6] D. Gallet, A. Waghray, P.N. Kumta, G. E. Blomgren, and M. Setter, in *Role of Ceramics in Advanced Electrochemical Systems*, P. N. Kumta, G. S. Rohrer, and U. Balachandran, Editors, Vol. 65, p. 177, American Ceramic Society Proceeding Series, Cincinnati, OH (1996).

[7] D. Gallet, A. Waghray, P. N. Kumta, G. E. Blomgren, and M. Setter, *J. Power Sources*, **72**, 91 (1998).

[8] W. Ebner, D. Fouchard, and L. Xie, *Solid State Ionics*, **69**, 238 (1994).

[9] R. V. Moshtev, P. Zlatilova, V. Manev, and A. Sato, *J. Power Sources*, **54**, 329 (1995).

[10] J. R. Dahn, U. von Sacken and C.A. Michal, *Solid State Ionics*, **44**, 87 (1990).

[11] S. Yamada, M. Fujiwara, and M. Kanda, *J. Power Sources*, **54**, 209 (1995).

[12] C. C. Chang and P. N. Kumta, *J. Power Sources*, **75**, 44 (1998).

[13] C. C. Chang, N. Scarr, and P. N. Kumta, *Solid State Ionics*, **112**, 329 (1998).

[14] D. Chen, X. Gao, D. Dollimore, Thermochim. Acta, 215 (1993) 65.

[15] D. Chen, X. Gao, D. Dollimore, Thermochim. Acta, 215 (1993) 109.

[16] X. Gao, D. Chen, D. Dollimore, Thermochim. Acta, 223 (1993) 75.

[17] D. Chen, D. Dollimore, J. Thermal Analysis, 44 (1995) 1001.

[18] D. Dollimore, T.A. Evans, Y.F. Lee, F.W. Wilburn, Thermochim. Acta, 188 (1991) 77.

[19] D. Dollimore, T.A. Evans, Y.F. Lee, G.P. Pee, F.W. Wilburn, Thermochim. Acta, 196 (1992) 225.

[20] D. Dollimore, T.A. Evans, Y.F. Lee, F.W. Wilburn, Thermochim. Acta, 198 (1992) 249.

[21] Jan N. Reimers, W. Li and J.R. Dahn, Physical Review B, Vol. 47, No. 14. 8486.

[22] R.V. Moshtev, P. Zlatilova, V. Manev, Atsushi Sato, Journal of Power Sources 54 (1995) 329.

[23] Shuji Yamada, Masashi Fujiwara, Motoya Kanda, Journal of Power Sources, 54 (1995) 209.

PHASE EVOLUTION AS A FUNCTION OF SYNTHESIS TEMPERATURE IN THE $Li_yMn_{3-y}O_{4+\delta}$ (0.7 ≤ y ≤ 1.33) SYSTEM

S. Choi and A. Manthiram
Texas Materials Institute, ETC 9.104
The University of Texas at Austin
Austin, TX 78712

ABSTRACT

$Li_yMn_{3-y}O_{4+\delta}$ (0.7 ≤ y ≤ 1.33) oxides have been synthesized by various procedures: an oxidation of aqueous Mn^{2+} with hydrogen peroxide or lithium peroxide in presence of lithium carbonate or lithium hydroxide and a sol-gel method followed by firing at 300 ≤ T ≤ 800 °C, and direct solid state reactions. An analysis of the phases formed as a function of firing temperature indicates that the spinel phase is formed only for a narrow range of 1.05 ≤ y ≤ 1.25 for the entire firing temperature range 300 ≤ T ≤ 800 °C. For example, Mn_2O_3 impurity phase is formed invariably for intermediate firing temperatures T ≈ 500 °C in the $LiMn_2O_{4+\delta}$ system (y = 1) irrespective of the synthesis procedure. The synthesis procedures have a minor influence on the phase evolution and carbon-containing raw materials generally lead to a larger amount of Mn_2O_3 impurity.

INTRODUCTION

Lithium-ion cells have become attractive power sources for portable electronic devices as they have higher energy density compared to other rechargeable systems. Layered $LiCoO_2$ serves as the cathode currently in commercial lithium-ion cells, but Co is expensive and toxic. There is enormous interest to develop alternate cathodes and manganese oxides are being pursued intensively since Mn is inexpensive and environmentally benign. Among the manganese oxides, the $LiMn_2O_4$ spinel is the most widely investigated material. $LiMn_2O_4$ exhibits two plateaus, one at 4 V and the other at 3 V, in the voltage versus lithium content curve, which correspond to the extraction/insertion of lithium from/into the 8a tetrahedral and 16c octahedral sites of the spinel lattice respectively.[1] The 3 V region involves a cubic to tetragonal transition caused by Jahn-Teller distortion, which is believed to be one of the reasons for the observed capacity fade in the cubic 4 V region; *i.e.* tetragonal grains formed on the surface of the cathodes in the 4 V region under the conditions of high rate or nonequilibrium are believed to cause capacity fade.[2]

To overcome these difficulties, several nonspinel manganese oxide such as the layered $LiMnO_2$,[2] orthorhombic $LiMnO_2$,[3] and amorphous manganese oxides[4] have been investigated. However, most of the nonspinel manganese oxides tend to transform to the spinel structure during the electrochemical cycling while the amorphous manganese oxides exhibit a continuous decrease in voltage during discharge. These considerations make the spinel manganese oxides the most viable candidates if the lattice distortion problems could be suppressed. One way to suppress the problems of Jahn-Teller distortion in spinel manganese oxides is to increase the oxidation state of manganese to 4+. In this regard, the Mn^{4+} spinel oxides $Li_4Mn_5O_{12}$ (i.e. $Li_{1+x}Mn_{2-x}O_4$ with $x = 0.33$) and $Li_2Mn_4O_9$ (i.e. $LiMn_2O_{4+\delta}$ with $\delta = 0.5$) are appealing.[5-8] However, both $Li_4Mn_5O_{12}$ and $Li_2Mn_4O_9$ are metastable and they tend to disproportionate to give $LiMn_2O_4$ spinel at higher temperatures $T > 500\ °C$. Obviously, these phases need to be accessed by low temperature synthesis procedures. Recently, our group[9] showed that $Li_4Mn_5O_{12}$ with all Mn^{4+} can be synthesized by firing at 400-500 °C a precursor obtained by oxidizing Manganese (II) acetate with lithium peroxide in aqueous solutions. We present in this paper the synthesis of $Li_yMn_{3-y}O_{4+\delta}$ ($0.7 \leq y \leq 1.33$) spinel oxides having a wide range of Li/Mn ratio by solution-based synthesis procedures. In addition, we investigate the evolution of phases as a function of synthesis temperature for the entire range of $0.7 \leq y \leq 1.33$. To our knowledge, a systematic study of the phase evolution with firing temperature (300 - 800 °C) for the $Li_yMn_{3-y}O_{4+\delta}$ system is not available in the literature.

EXPERIMENTAL

$Li_yMn_{3-y}O_{4+\delta}$ ($0.77 \leq y \leq 1.33$) samples were synthesized by oxidizing Mn^{2+} in aqueous solution in presence of excess Li^+ salts. Specific amounts of aqueous solutions of lithium carbonate or lithium hydroxide and hydrogen peroxide or lithium peroxide were first mixed together. The mixture was then added immediately to 20 ml of 0.25 M manganese acetate solution that was kept under constant stirring on a magnetic stirrer. After stirring the mixture for about 5 minutes, the $Li_xMn_yO_z \cdot nH_2O$ precipitate formed was filtered and dried. The precursors thus obtained were then fired in air at various temperatures $300 \leq T \leq 800\ °C$ with heating and cooling rates of, respectively, 5 and 2 °C/min. In order to evaluate the possible influence of synthesis procedures on the phase evolution, the $y = 1$ member was also synthesized by solid state reaction between Li_2CO_3 or $LiOH \cdot H_2O$ and MnO_2 or $MnCO_3$ at 400-800 °C in air and by a sol-gel procedure. In the sol-gel procedure, required amounts of manganese acetate and lithium carbonate were dissolved in acetic acid and refluxed for 1 h. Small amounts of water and hydrogen peroxide were then added, and the mixture was refluxed until

Table I. Synthesis procedures for $Li_yMn_{3-y}O_{4+\delta}$

Number	Synthesis method	Reactants
1	oxidation reaction	320 mL of various concentrations of Li_2CO_3 + 60 mL of 30% H_2O_2 + 20 mL of 0.25 M $Mn(CH_3COO)_2 \cdot 4H_2O$
2	oxidation reaction	160 mL of various concentrations of $LiOH \cdot H_2O$ + 80 mL of 0.25 M Li_2O_2 + 20 mL of 0.25 M $Mn(CH_3COO)_2 \cdot 4H_2O$
3	oxidation reaction	160 mL of 0.25 M $LiOH \cdot H_2O$ + 60 mL of 30 % H_2O_2 + 20 mL of 0.25M $Mn(CH_3COO)_2 \cdot 4H_2O$ (y = 1 sample)
4	sol-gel processing	Li_2CO_3 + 4 $Mn(CH_3COO)_2 \cdot 4H_2O$ (y = 1 sample)
5	solid state reaction	Li_2CO_3 + 4 MnO_2 (y = 1 sample)
6	solid state reaction	$LiOH \cdot H_2O$ + 4 MnO_2 (y = 1 sample)
7	solid state reaction	Li_2CO_3 + 4 $MnCO_3$ (y = 1 sample)

a clear solution was formed. The solution was then heated on a hot plate to form a transparent gel, which was then fired in air at 300-800 °C. The various synthesis methods used are summarized in Table I. In all the procedures, a firing duration of 3 days was used for $300 \leq T \leq 500$ °C and 1 day was used for $600 \leq T \leq 800$ °C unless otherwise specified.

The fired samples were characterized by X-ray powder diffraction. Lithium and manganese contents were determined by atomic absorption spectroscopy. Oxygen contents were determined by a redox titration involving the dissolution of the sample in a known excess of sodium oxalate solution in presence of dilute sulfuric acid and titrating the unreacted sodium oxalate with potassium permanganate.

RESULTS AND DISCUSSION

Phase analysis results of the $Li_yMn_{3-y}O_{4+\delta}$ ($0.77 \leq y \leq 1.33$) samples obtained by synthesis procedure 1 in Table I are presented in Fig. 1. The data show that spinel phases are formed for the entire range of $0.77 \leq y \leq 1.33$ for firing temperatures $T \leq 300$ °C. However, the samples fired at $T \leq 300$ °C are poorly crystalline as indicated by the broad X-ray reflections in Fig. 2. As the firing temperature increases, the metastable spinel phases tend to disproportionate for most of the y values in $Li_yMn_{3-y}O_{4+\delta}$ excepting for y = 1.13. Only the y = 1.13 composition gives the spinel phase for the entire firing temperature range $300 \leq T \leq 800$ °C. For $1 \leq y < 1.13$, single phase spinel is formed for lower (400 °C) and higher (800 °C) firing temperatures, but Mn_2O_3 impurity is formed for intermediate firing temperature $T \approx 500$ °C. The formation of Mn_2O_3 impurity for intermediate firing temperature $450 \leq T \leq 700$ °C is evident in the X-ray diffraction patterns given in Fig. 2 for the y = 1 member. For $1.13 < y \leq 1.33$, single phase spinel is formed only for lower (400 °C) firing temperatures, and Li_2MnO_3 impurity is formed for higher (800 °C) firing temperature. For some compositions with y > 1.13, Mn_2O_3 impurity is observed for intermediate firing temperatures. Though confusing, one possible reason for the formation of Mn_2O_3 impurity is the reduction caused by the use of carbon-containing precursors. It also implies that the samples may not be equilibrated.

In order to verify the possibility that the use of carbon-containing precursors may be responsible for the formation of the Mn_2O_3 impurity for intermediate firing temperatures, we synthesized the samples by using $LiOH \cdot H_2O$ instead of Li_2CO_3 in the oxidation procedure (synthesis method 2 in Table I). The phase

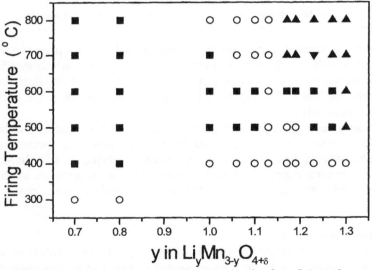

Fig. 1 Phase analysis of the products obtained after firing the precursors of $Li_yMn_{3-y}O_{4+\delta}$ ($0.7 \leq y \leq 1.33$) at various temperatures in air. The precursor was obtained by oxidizing manganese acetate solution with H_2O_2 in presence of Li_2CO_3. O: spinel; ■: spinel + Mn_2O_3; ▲: spinel + Li_2MnO_3; ▼: spinel + Mn_2O_3 + Li_2MnO_3.

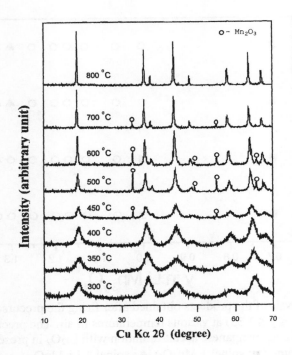

Fig. 2 X-ray diffraction patterns recorded after firing the precursor of sample 1 in Table I at various temperatures in air. The precursor was obtained by oxidizing manganese acetate solution with H_2O_2 in presence of Li_2CO_3. The unmarked reflections correspond to the spinel phase.

analysis results are summarized in Fig. 3. We see that single phase spinel is formed for the entire firing temperature range $500 \leq T \leq 800°C$ for a wider range of $1.05 \leq y \leq 1.25$ compared to that in Fig. 1. Also, the confusion of Mn_2O_3 impurity formation at intermediate firing temperature for $y > 1.13$ is no longer present. The use of carbon-free precursor $LiOH \cdot H_2O$ and Li_2O_2 seems to suppress the formation of Mn_2O_3 and favor the formation of spinel phase for a wider range of y. Thus the precursor seems to play an important role on the phase evolution.

In order to gain further understanding, we have also synthesized the $y = 1$ member by a few other procedures as listed in Table I (methods 3 to 7). It was found that Mn_2O_3 impurity was formed at intermediate temperatures in all cases regardless of the synthesis procedures although the amount of the impurity and formation temperature varied depending on the synthesis method. From the data of synthesis methods 1 to 4 in Table I, it was concluded that the use of carbon-

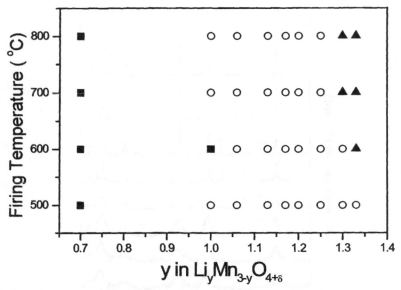

Fig. 3 Phase analysis of the products obtained after firing the precursors of $Li_yMn_{3-y}O_{4+\delta}$ (0.7 ≤ y ≤ 1.33) at various temperatures in air. The precursor was obtained by oxidizing manganese acetate solution with Li_2O_2 in presence of $LiOH \cdot H_2O$. O: spinel; ■: spinel + Mn_2O_3; ▲: spinel + Li_2MnO_3.

containing precursors such as Li_2CO_3 in the solution-based oxidation method or Li_2CO_3 and acetic acid in the sol-gel method leads to higher amount of Mn_2O_3 for intermediate firing temperatures compared to the use of carbon-free precursors; however, it should be noted that a carbon-containing raw material, manganese acetate, was used in all the experiments. It appears that the large amount of carbon present in the precursors tends to reduce the manganese ions during firing and favors the formation of Mn_2O_3 impurity.

In addition, samples obtained by direct solid state synthesis (methods 5 to 7 in Table I) also showed the formation of Mn_2O_3 impurity for intermediate firing temperatures 500 ≤ T ≤ 700 °C. This observation is in agreement with that found previously by Siapkas et al[10] in a direct solid state synthesis involving the reaction between Li_2CO_3 and MnO_2. However, the use of carbon-free precursors in the case of direct solid state synthesis (method 6 in Table I) does not lead to a suppression of the amount of Mn_2O_3 impurity. Among the three solid state syntheses (methods 5 to 7 in Table I), only the procedure that employs Li_2CO_3 and $MnCO_3$ as raw materials gave single phase spinel for a lower firing temperature T = 400 °C. The observation of single phase spinel at 400 °C for

Materials for Electrochemical Energy Conversion and Storage

synthesis method 7 is in agreement with that found by de Kock et al.[5] However, we find that firing for prolonged time (3 days) at 400 °C does not lead to the formation of impurity phases unlike that pointed out by de Kock et al.[5] In other solid state syntheses that employed MnO_2 (methods 5 and 6 in Table I), the firing temperature of 400 °C is, however, too low and the reaction appears to be incomplete as indicated by the presence of MnO_2 impurity at 400 °C.

It is clear that Mn_2O_3 impurity is formed for intermediate firing temperatures $T \approx 500$ °C in the system $LiMn_2O_{4+\delta}$ irrespective of the synthesis method. The data in Figs. 1 and 3 reveal that lithium-rich spinels $Li_{1+x}Mn_{2-x}O_{4+\delta}$ are more stable compared to $LiMn_2O_{4+\delta}$ for intermediate firing temperatures. The stability of lithium-rich spinels for intermediate firing temperatures could be due to a tendency to maximize the oxidation state of manganese. The formation of Li_2MnO_3 impurity at higher temperatures for $y > 1.13$ in Fig. 1 or $y > 1.25$ in Fig. 2 is due to a disproportionation of the lithium-rich spinels to the thermodynamically more stable $LiMn_2O_4$ spinel and a tendency to lower the oxidation state of manganese in the spinel phase. It should be noted that Li_2MnO_3 impurity is sometimes difficult to detect in the X-ray diffraction patterns as the strong reflections of Li_2MnO_3 overlap with the reflections of the spinel phase. However, Li_2MnO_3 is red in color while the spinel phases are black in color. So in many instances, the formation of Li_2MnO_3 was inferred from the red particles present in the sample. The data in Figs. 1 and 3 in fact refer to the observation of Li_2MnO_3 either by color or by X-ray diffraction.

Finally, wet-chemical analysis data show that the oxidation state of manganese in the spinel phase decreases with increasing firing temperature. For example, $LiMn_2O_{4+\delta}$ ($y = 1$ member) has a manganese oxidation state of 3.82+ and 3.50+, respectively, for the firing temperatures of 300 and 800 °C. Thus the Mn^{4+} defect spinel "$Li_2Mn_4O_9$" with $\delta = 0.5$ could not be accessed even at the lowest firing temperature of 300 °C although the formation of $Li_2Mn_4O_9$ has been claimed from neutron diffraction data.[5]

CONCLUSIONS

$Li_yMn_{3-y}O_{4+\delta}$ ($0.7 \leq y \leq 1.33$) oxides have been synthesized by various solution-based and solid state methods. Phase analysis as a function synthesis temperature reveals that the spinel phase is formed only for a narrow range of $1.05 \leq y \leq 1.25$ for the entire firing temperature range $300 \leq T \leq 800$ °C. Single phase $LiMn_2O_{4+\delta}$ spinel is formed either at lower temperatures or at higher temperatures and Mn_2O_3 impurity phase is formed at intermediate firing temperatures $T \approx 500$ °C. This finding implies that the lithium-rich spinel oxides $Li_{1+x}Mn_{2-x}O_4$ are more stable compared to the $LiMn_2O_4$ spinel at the intermediate firing temperatures.

ACKNOWLEDGMENT

Financial support by the Welch Foundation Grant F-1254 is gratefully acknowledged.

REFERENCES

[1] M. M. Thackeray, Y. Shao-Horn, A. J. Kahaian, K. D. Kepler, E. Skinner, J. T. Vaughey, and S. A. Hackney, "Structural Fatigue in Spinel electrodes in High Voltage(4V) Li/Li$_x$Mn$_2$O$_4$ Cells," *Electrochemical and Solid-State Letters*, **1**, 7 (1998).

[2] R. J. Gummow, D. C. Liles, and M. M. Thackeray, "Lithium Extraction from Orthorhombic Lithium Manganese Oxide and the Phase Transformation to Spinel," *Materials Research Bulletin*, **28**, 1249 (1993).

[3] A. R. Armstrong, and P. G. Bruce, "Synthesis of Layered LiMnO$_2$ as an Electrode for Rechargeable Lithium Batteries," *Nature* **381**, 499 (1996).

[4] J. Kim, and A. Manthiram, "A Manganese Oxyiodide Cathode for Rechargeable Lithium Batteries," *Nature* **390**, 265 (1997).

[5] A. de Kock, M. H. Rossouw. L. A. de Picciotto, M. M. Thackeray, W. I. F. David, and M. R. Ibberson, "Defect Spinels in the System Li$_2$O.yMnO$_2$ (y > 2.5): A Neutron-Diffraction Study and Electrochemical Characterization of Li$_2$Mn$_4$O$_9$," *Materials Research Bulletin*, **25**, 657 (1990).

[6] M. M. Thackeray, A. De Kock, M. H. Rossouw, D. C. Liles, D. Hoge, and R. Bittihn, "Spinel Electrodes from the Li-Mn-O System for Rechargeable Lithium Battery Application," *Journal of the Electrochemical Society*, **139**, 363 (1992).

[7] M. M. Thackeray, and A. de Kock, "Synthesis and Structural Characterization of Defect Spinels in the Lithium-Manganese-Oxide System," *Materials Research Bulletin*, **28**, 1041 (1993).

[8] T. Takada, H. Hayakawa, and E. Akiba, "Preparation and Crystal Structure Refinement of Li$_4$Mn$_5$O$_{12}$ by the Rietveld Method," *Journal of Solid State Chemistry*, **115**, 420 (1995).

[9] J. Kim, and A. Manthiram, "Synthesis of Spinel Li$_4$Mn$_5$O$_{12}$ Cathodes by Oxidation Reaction in Solution," *Journal of the Electrochemical Society*. **145**, L53 (1998).

[10] D. I. Siapkas, I. Samaras, C. L. Mitsas, E. Hatzikraniotis, T. Zobra, D. Terzidis, G. Moumouzias, S. Kokkou, A. Zouboulis, and K. M. Paraskevopoulos, "Characterization of Optimized Cathodes Prepared from Synthesized Spinel LiMn$_2$O$_4$ for Li-Ion Battery Applications," *Electrochemical Society Proceedings*, **97-18**, 199 (1997).

Si/TiN Nanocomposite Anodes by High-Energy Mechanical Milling

Il-seok Kim[a] and Prashant N. Kumta[a]

[a]Carnegie Mellon University, Pittsburgh, Pennsylvania 15213

G. E. Blomgren[a,*]

*Blomgren consulting Services Ltd., 1554 Clarence Avenue, Lakewood, Ohio 44107

ABSTRACT

Nanocomposites containing silicon and titanium nitride were synthesized by high-energy mechanical milling (HEMM). The process results in very fine Si particles distributed homogeneously inside the TiN matrix. The Si/TiN nanocomposites synthesized using different experimental conditions were evaluated for their electrochemical properties. Results indicate that Si in the composite alloys and de-alloys with lithium during cycling while TiN remains inactive providing the desired structural stability. The composite containing 33.3 mol% Si obtained after 12 h milling exhibited a capacity of $\approx$ 300 mAh/g, reflecting its promising nature. Preliminary cycling data show good capacity retention indicative of good phase and micro-structural stability as verified by XRD and SEM analyses.

INTRODUCTION

Graphite has been the customary anode material for lithium ion batteries with a theoretical capacity of 372 mAh/g or volumetric capacity of 830 Ah/L.[1] During last few years however, there has been some activity focused on identifying alternative anode materials. Most recent developments to date include tin oxide based composites and nanocomposites of intermetallics, comprising electrochemically 'active-inactive' phases.[2-6] Intermetallic compounds containing lithium have also been studied extensively as anode materials, due to their higher gravimetric/volumetric capacities in comparison to graphite. However, a major problem with lithium alloys as anodes is the very large change in volume during charge/discharge resulting in cracking or crumbling of the anodes causing loss in capacity during cycling.[1]

In order to preserve and stabilize the original morphological state of the anode and thereby attain good electrochemical properties, various material systems have been analyzed to minimize the mechanical stress induced by the large ensuing volumetric changes of the active phase. Most of the current studies on anode materials other than carbon have focused on creating a composite microstructure comprising an inactive host matrix containing a finely dispersed active phase that remains interconnected.[2-6] Although these systems are promising, there are problems related to either irreversible loss, capacity and/or cyclability. As a result, there is a need for further improving the demonstrated concept of the 'active-inactive' composites. In order to improve this concept further, thdre is a need to identify a system containing elements that exhibit high discharge capacity with repect to lithium. Additionally, matrix systems would need to be identified that exhibit good mechanical strength and electrical conductivity. Above all, they should also be primarily electrochemically inactive against lithium. In this paper, we demonstrarate that the Si-TiN system has considerable potential for use as anodes in Li-ion battery applications.

The purpose of this paper is therefore to explore the use of silicon/titanium nitride (Si/TiN) composite as an anode material synthesized by exploiting the technique of high-energy mechanical milling (HEMM). TiN is a well-known material and it has several applications.[7-9] Due to its good electrical conductivity, the high surface area form of TiN is also recently attracting interest for supercapacitor applications.[10, 11] However, to the best of our knowledge, there have been no reports to date on its use as an electrode in lithium ion batteries. The basic objective of this study is therefore to exploit the use of TiN as an electrochemically inactive matrix in the presence of an electrochemically active phase of Si. This is because of the well-known electrical and mechanical properties of TiN such as electrical conductivity, mechanical strength, combined with electrochemical inertness to Li (potential range $0.02 \rightarrow 1.2$ V)[12] and chemical inertness to both Li and Si. The inert electrochemical and chemical characteristics of TiN to Si can only be exploited if a suitable process for synthesizing such composites can be identified. In this context, the technique of HEMM is indeed very promising due to its proven ability to generate amorphous, metastable and nanophase structures.[13] The present paper therefore focuses on using HEMM for generating novel nanocomposites in the Si-TiN system for use as anodes in Li-ion rechargeable batteries. Experimental studies and results of the structural and electrochemical analyses are presented indicating its promising nature.

EXPERIMENTAL

Nanocomposites of Si and TiN were prepared using a SPEX-8000 high-energy mechanical mill. Commercial elemental powder of Si (Aldrich, 99 %) and TiN (Aldrich, 99 %)) were used as starting materials. Stoichiometric amounts of both powders were weighed and loaded into a hardened steel vial containing hardened steel balls. All the handling and batching of the powders in the vial prior to milling were conducted inside the glovebox (VAC Atmospheres, Hawthorn, CA). The vial was also firmly sealed to prevent and minimize any oxidation of the starting materials.

In order to evaluate the electrochemical characteristics, electrodes were fabricated using the as-milled powder by mixing the composition of 87.1 wt% active powder and 7.3 wt% acetylene black. A solution containing 5.6 wt% polyvinylidene fluoride (PVDF) in 1-methyl-2-pyrrolidinone (NMP) was added to the mixture. The as-prepared solution was coated onto a Cu foil. A hockey puck cell design was used employing lithium foil as an anode and 1 M LiPF$_6$ in EC/DMC (2:1) as the electrolyte. All the batteries tested in this study were cycled for 20 cycles in the voltage range from 0.02~1.2 V employing a current density of 0.25 mA/cm^2 and one-minute rest period between the charge/discharge cycles using a potentiostat (Arbin electrochemical instrument). The phases present in the as-milled powders and the cycled electrode were analyzed using x-ray diffraction (Rigaku, θ/θ diffractometer), while the microstructure and chemical composition of the electrode was examined using a scanning electron microscope (Philips XL30, equipped with EDX).

RESULT AND DISCUSSIOIN

Preliminary electochemical trial experiments conducted on Si/TiN composites comprising a molar ratio of Si:TiN = 1:2 (i.e. 33.3 mol% Si) exhibited the best electrochemical properties. Hence, the rest of the work focused on this composition of the composite. In order to analyze the phases present after milling, x-ray diffraction was conducted on the as-milled powders obtained after milling for various time periods (see Fig 1(a)). All the peaks in the patterns correspond to TiN and the broad nature of the peaks suggests the nanocrystalline nature of the TiN powder. The non-observance of any Si related peak in XRD indicates that Si exists in a nanocrystalline-amorphous form well dispersed inside the powder even after milling for only 6 h.

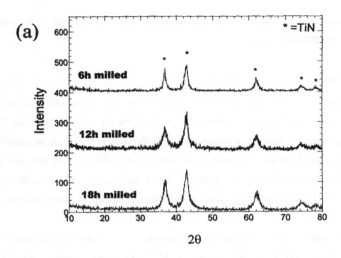

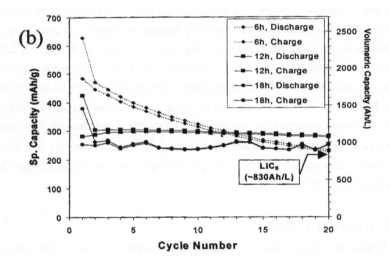

Figure 1. (a) X-ray diffraction patterns of Si:TiN=1:2 composites milled for 6 h, 12 h and 18 h, respectively. (b) Capacity as a function of cycle number for Si/TiN nanocomposites obtained after milling for 6 h, 12 h and 18 h each.

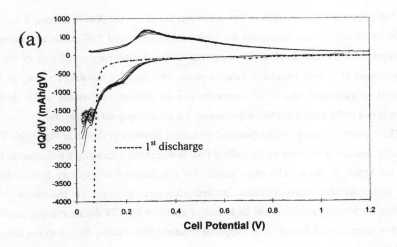

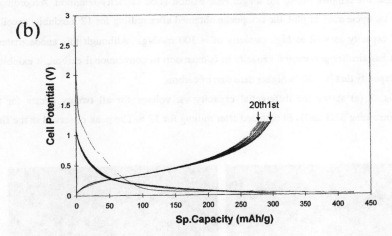

Figure 2. (a) Differential capacity vs. cell potential curves, (b) cell potential vs. specific capacity curves for the first twenty cycles of the 12 h milled Si/TiN composite with Si:TiN = 1:2 molar ratio obtained after milling for 12 h.

This suggests that the HEMM process provides enough energy to generate the nanocomposite powder of Si and TiN. It is not clear based on these results whether Si reacts with TiN to form any Si-N type bonds. More extensive structural characterization using MASS-NMR may be necessary to identify the existence of these types of environments. Based on the XRD patterns, it can be convincingly mentioned that the composites are composed of nanosized TiN matrix containing a uniform dispersion of Si regardless of the composition. Due to the very small size of the Si crystals, the impact of volume expansion induced phase transitions of the active phase, on the inactive matrix is minimized Hence the composite can be expected to remain stable during cycling. This is one of the most important requirements for achieving good cyclability.

The capacity of the electrodes prepared with these powders is shown in Fig. 1 (b). The overall capacity appears to decrease as the milling time is increased, indicating a reduction in the amount of the active Si phase. The exact reason for the decrease in capacity is not clear, warranting further detailed characterization studies. However, it could be speculated with conviction that the Si nanoparticles could be buried or enclosed by TiN during milling, thereby preventing their reaction with lithium. The composite obtained after milling for 6 h shows fade in capacity, while the samples milled for longer time exhibit good capacity retention. According to the capacity vs. cycle number plot, the composite obtained after milling for 12 h exhibits excellent retention of capacity as well as high capacity of $\approx$ 300 mAh/g. Although this anode material composition has smaller gravimetric capacity in comparison to conventional carbon, it exhibits a volumetric capacity that is ~ 30 % higher than that of carbon.

Fig. 2 (a) shows the differential capacity vs. voltage for all twenty cycles for the composite containing 33.3 mol% Si obtained after milling for 12 h. The peak observed in the first

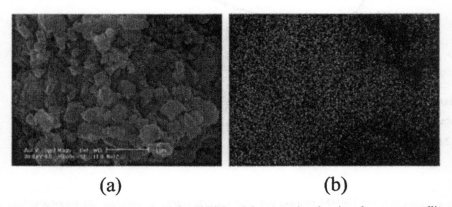

(a)   (b)

Figure 3. (a) SEM micrograph of the Si:TiN = 1:2 composite showing the nanocrystalline particles. (b) Chemical map of Si using EDX for the Si/TiN composite with Si:TiN = 1:2 molar ratio obtained after milling for 12 h. (Both images are taken at the same scale.)

(a)

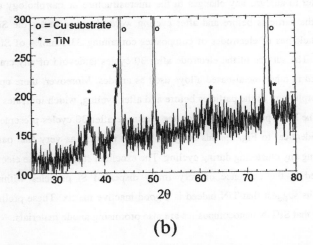

(b)

Figure 4. (a) SEM micrographs of the electrode before and after 30 cycles, (b) XRD pattern of the electrode after 30 cycles prepared with the Si/TiN composite corresponding to Si:TiN = 1:2 molar ratio obtained after milling for 12 h.

cycle is bigger and has a different shape compared to subsequent cycles, suggesting the expected irreversible loss in capacity. Dahn et al. reported in their recent study on Sn_2Fe that peak sharpening in the differential capacity curve corresponds to aggregation of active Sn particles during cycling.[14] However, no peak sharpening occurred even after 20 cycles in our electrodes, indicating the stability of the microstructure without aggregation of Si particles. The plot in Fig. 2

(b) exhibits the relation between cell potential and capacity. The curves show a smooth plateau in the low voltage range as well as good capacity retention. The reason for the first cycle irreversible loss shown in the plot is still unclear and needs more detailed study but one of the more obvious reasons could be the formation of Li-containing passivation layer.

The SEM micrograph and the energy dispersive elemental x-ray (EDX) map for silicon obtained on the 12 h milled powder containing 33.3 mol% Si are shown in Fig. 3. The particles are agglomerated although they are extremely small in the range of 100~500 nm. All these particles are therefore true composites containing an intimate mixture of Si and TiN according to the EDX analysis. Thus no crystalline Si peaks are observed in the XRD patterns. The EDX results also show the presence of iron ($\approx$ 3.3 %) in the as-milled powder, suggesting that iron is incorporated from the vial or the balls used during milling.

In order to analyze any changes in the microstructure or morphology of the particles during cycling, the particles before and after cycling were observed under the SEM. Fig. 4 (a) shows the morphologies of electrodes of composites containing 33.3 mol% of Si obtained after milling for 12 h. The surface of the electrode after 30 cycles is devoid of any cracks, which are typically observed in other metal-based alloys used as anodes. Moreover, there appears to be no change in the morphology of the particles before and after cycling, which indicates the stability of the composite. The XRD pattern obtained on the electrode after 30 cycles presented in Fig. 4 (b) shows that it is identical to Fig. 1. This suggests that Si remains as very fine particles existing without undergoing any clustering during cycling. The excellent stability of the electrode therefore may be attributed to the existence of very finely dispersed Si particles within TiN. These encouraging results suggest that TiN indeed is a good inactive matrix. These preliminary studies also clearly show that Si/TiN nanocomposites are also promising anode materials.

CONCLUSIONS

Nanostructured Si/TiN composites can be produced by HEMM and the powders prepared by this method are composed of amorphous Si and nanosized TiN. As the amount of active Si is reduced when the milling time is increased, a reduction in the initial specific capacity was observed. The milling time appears to control the amount of active Si exposed to lithium since prolonged milling can lead to an increase in the inactive portion of Si. The electrode containing 33.3 mol% Si, milled for 12 h shows good capacity (~300 mAh/g or ~1100 Ah/L) with little fade (~0.36 % /cycle). The as-milled powder consists of agglomerates of nanosized particles, while a single particle itself is essentially a nanocomposite of Si and TiN as indicated by EDX results. The electrode structure is also very stable during cycling because no cracking/crumbling and/or obvious clustering of Si was observed after 30 cycles. These initial results suggest that the material system certainly appears to be promising as an anode material although further optimization studies need to be conducted in order to demonstrate its optimum properties. Detailed structural and electrochemical studies are currently in progress and will be reported in subsequent publications.

ACKNOWLEDGEMENTS

P. N. Kumta and Il-seok Kim would like to acknowledge the support of NSF (CTS Grant 9700343). P. N. Kumta, Il-seok Kim and G. E. Blomgren would also like to thank the financial support of ONR (Grant N00014-00-1-0516). Changs Ascending (Taiwan) and Pittsburgh Plate Glass (Pittsburgh) are also acknowledged for providing partial financial support.

REFERENCES

[1] R. A. Huggins, *Solid State Ionics*, **113-115** (1998) 57-67.

[2] K. D. Kepler, J. T. Vaughey, and M. M. Thakeray, *Electrochem. Solid-State Lett.*, **2** (1999) 307.

[3] O. Mao and J. R. Dahn, *J. Electrochem. Soc.*, **146** (1999) 423.

[4] Y. Idota, A. Matsufuji, Y. Maekawa, and T. Miyasaki, *Science*, **276** (1997) 1395.

[5] M. Winter and J. O. Besenhard, *Electrochim. Acta*, **45** (1999) 31.

[6] H. Kim, J. Choi, H. J. Sohn, and T. Kang, *J. Electrochem. Soc.*, **146** (1999) 440.

[7] H. Zheng, K. Oka, and J. D. Mackenzie, *Mat. Res. Soc. Symp. Proc.*, **271** (1992) 893.

[8] T. Granziani and A. Bellosi, *J. Mater. Sci. Lett.*, **14** (1995) 1078.

[9] K. Kamiya and T. Nishijima, *J. Am. Ceram. Soc.*, **73** (1990) 2750.

[10] C. F. Windisch, Jr., J. W. Virden, S. H. Elder, J. Liu, and M. H. Engelhard, *J. Electrochem. Soc.*, **145** (1998) 1211.

[11] M. R. Wixom, D. J. Tarnowski, J. M. Parker, J. Q. Lee, P. L. Chen, I. Song, and L. T. Thompson, *Mat. Res. Soc. Symp. Proc.*, **496** (1997) 643.

[12] I. S. Kim and P. N. Kumta, unpublished data (1999).

[13] E. Gaffet, F. Bernard. J. C. Niepce, F. Charlot, C. Gras, G. L. Caer, J. L. Guichard, P. Delcroix, a. Mocellin, and O. Tillement, *J. Mater. Chem.*, **9** (1999) 305.

[14] O. Mao, R. A. Dunlap, and J. R. Dahn, *J. Electrochem. Soc.*, **146** (1999) 405.

Papers from 2001 Meeting

Gas Separation Membranes

OXYGEN PERMEATION THROUGH MIXED-CONDUCTING PEROVSKITE OXIDE MEMBRANES

H.J.M. Bouwmeester and L.M. van der Haar
Laboratory for Inorganic Materials Science
Faculty of Chemical Technology & MESA$^+$ Research Institute
University of Twente
PO Box 217, 7500 AE Enschede, the Netherlands

ABSTRACT

In this paper, we summarise data of thermodynamics and transport of oxygen in phases $La_{1-x}Sr_xCoO_{3-\delta}$ obtained from oxygen coulometric titration and conductivity relaxation measurements. Emphasis is on recent results from oxygen permeation measurements. The results clearly demonstrate that the usual assumption of randomly distributed non-interacting oxygen vacancies in these perovskite compositions is an oversimplified picture.

INTRODUCTION

Mixed oxygen-ionic and electronic conducting perovskite oxide membranes have attracted major research interest in recent years due to their possible use in various oxygen delivery applications. Provided that they can be prepared free of cracks or connected-through porosity, the ceramic membranes facilitate separation of oxygen from an air supply with infinite selectivity [1]. The membranes offer considerable promise for use in the production of oxygen or oxygen-enriched gas mixtures. Closely allied to this application is their use in membrane reactors for hydrocarbon conversion, combining oxygen separation and catalytic reaction into a single step, for example, converting methane to synthesis gas.

Presently, perovskite oxides $La_{1-x}A_xCo_{1-y}Fe_yO_{3-\delta}$ (A = Sr, Ba) have been studied intensively as candidate membrane materials. Taking $LaCoO_{3-\delta}$ as the parent compound, the partial substitution of A and B-site cations in the ABO_3 perovskite structure by lower valent dopants is charge compensated by a dual mechanism, involving the creation of oxygen vacancies and electronic holes. The partial occupation of the oxygen sublattice results in materials with high ionic conductivity, typically above 700°C. Imposing an oxygen partial pressure differential across the membrane drives oxygen from the high partial pressure side to the low partial pressure side. In contrast with fuel cells and oxygen pumps the mixed conducting membranes operate without the need of electrodes and external circuitry. The presence of high electronic conductivity thereby acts as an internal short-circuit for the return path of electrons to counteract the flux of oxygen anions. To maintain structural and chemical integrity under the very reducing conditions, under which reactors for synthesis gas operate, additional B-site doping (Cr, Mn, Ga) may be applied.

In this paper, we summarise results obtained from thermodynamic and transport measurements on phases $La_{1-x}Sr_xCoO_{3-\delta}$ using combined oxygen coulometric titration, conductivity relaxation and oxygen permeation measurements.

HIGH-TEMPERATURE OXYGEN COULOMETRIC TITRATION

Solid state electrochemical cells employing oxide ion conducting electrolytes are very attractive for the study of thermodynamic properties of an oxide. The equilibrium chemical potential, partial molar energy and entropy of oxygen in perovskites $La_{1-x}Sr_xCoO_{3-\delta}$ were measured as a function of δ and x by high-temperature coulometric titration (650-950°C), utilising yttrium-stabilised zirconia as the solid electrolyte [2,3]. Figure 1a shows the equilibrium oxygen chemical potential μ_{O_2} as a function of oxygen nonstoichiometry δ for compositions $La_{1-x}Sr_xCoO_{3-\delta}$ with $x = 0.2$, 0.4 and 0.7. The data are replotted in Fig. 1b, showing that the curves for the different compositions almost coincide. An almost linear decrease of μ_{O_2} is observed with increasing the net electron concentration $(2\delta-x)$. To a first approximation the observed relationship can be fitted with

$$\mu_{O_2} \, [kJ/mole] = -340 - 345 \times (2\delta - x) \tag{1}$$

The data can be accounted for by assuming that the $(2\delta-x)$ electrons induced by vacancy formation or reduction of the strontium dopant level are filling up states in a wide electron band. This is consistent with the observed metallic behaviour of electrons in phases $La_{1-x}Sr_xCoO_{3-\delta}$. Accordingly, changes in the chemical potential of oxygen with variation in temperature, oxygen partial pressure and strontium content reflect concomi-

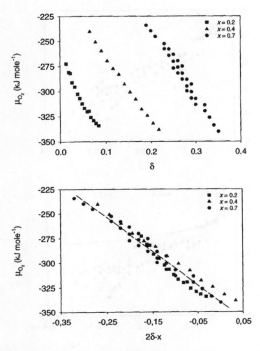

Fig. 1. Plot of the oxygen chemical potential μ_{O_2} as a function of a) δ and b) $(2\delta\text{-}x)$ for $La_{1\text{-}x}Sr_xCoO_{3\text{-}\delta}$ ($x = 0.2$, 0.4 and 0.7). The drawn line indicates the fit of the data to Eq. (1).

tant shifts in the relative position of the Fermi-level with a slope determined by the local density of states.

The above explanation for the observed relationship between μ_{O_2} and electron occupancy also requires that variations in the chemical potential of oxygen vacancies are only small. Taking the entropy of the electrons to be zero, estimates can be obtained for the partial molar enthalpy and entropy of oxygen vacancies from the temperature dependence of μ_{O_2} at constant oxygen stoichiometry (Fig. 2). For low values of δ and at the highest temperatures, the oxygen vacancies in $La_{1\text{-}x}Sr_xCoO_{3\text{-}\delta}$ turn out to be randomly distributed among equivalent oxygen sites. By increasing δ or lowering temperature additional ionic contributions to the partial energy and

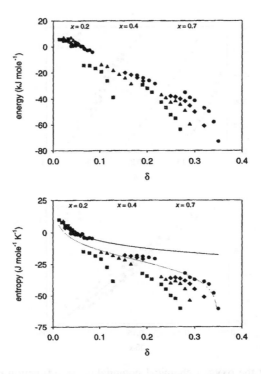

Fig. 2. Plot of a) the energy and b) the entropy associated with vacancy formation in $La_{1-x}Sr_xCoO_{3-\delta}$ as a function of δ for $x = 0.2$ at 800°C (■), 850°C (▲), 900°C (◆) and 950°C (●), and both $x = 0.4$ and 0.7 at 688°C (■), 763°C (▲), 838°C (◆) and 913°C (●). The solid line represents the entropy of randomly distributed oxygen vacancies in a lattice with 3 oxygen sites per unit cell. Similarly, the dashed line corresponds with the random distribution among 0.35 sites per unit cell.

entropy arise, but these cancel in the chemical potential of oxygen vacancies. The results are indicative for equilibrium on the oxygen sublattice. Concordant with the data from high temperature oxygen-17 NMR on related compounds [4], this is attributed to the presence of nanosized domains of ordered oxygen vacancies, co-existing with regions in which vacancies are randomly distributed.

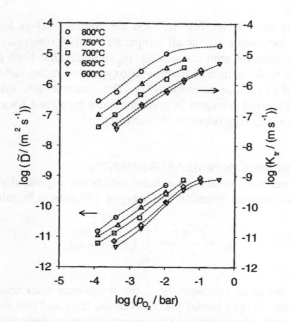

Fig. 3. The chemical diffusion coefficient $\tilde{D}$ and surface exchange coefficient K_{tr} as a function of oxygen partial pressure for $La_{0.3}Sr_{0.7}CoO_{3-\delta}$. The lines are an aid to link the data and have no theoretical significance.

OXYGEN CONDUCTIVITY RELAXATION

Chemical diffusivity and surface oxygen exchange of $La_{1-x}Sr_xCoO_{3-\delta}$ ($x = 0.2$, 0.5 and 0.7) were studied using the conductivity relaxation technique [5]. This technique involves measurement of the time variation of the electrical conductivity of a sample after a stepwise change in the ambient oxygen partial pressure. The relaxation data are fitted to theoretical equations, using the chemical diffusion coefficient $\tilde{D}$ and surface exchange rate K_{tr} as fitting parameters. Measurements were performed in the range of temperature 600-850°C and oxygen partial pressure 10^{-4}-1 bar. Typical results as obtained for $La_{0.3}Sr_{0.7}CoO_{3-\delta}$ are shown in Fig. 3. Similar results were obtained for the two other compositions. The expected behaviour for a random distribution of oxygen vacancies in the perovskite lattice is that of $\tilde{D}$ increasing with the extent of oxygen nonstoichiometry, *i.e.* increasing with decreasing p_{O_2}. On the contrary, $\tilde{D}$ is found to decrease with decreasing p_{O_2} at all temperatures. The observation that $\tilde{D}$ and K_{tr} both decrease with decreasing p_{O_2} suggests that both parameters are correlated with the mobile fraction of oxygen vacancies rather than the full extent of oxygen nonstoichiometry. The characteristic thickness L_c, below which overall transport is predominantly governed by surface exchange, is found to vary between 50 and 150 μm.

OXYGEN PERMEATION MEASUREMENTS

Oxygen permeation through a dense membrane is generally described by Wagner's equation, assuming bulk oxygen diffusion to be rate limiting,

$$j_{O_2} = -\frac{RT}{4^2 F^2 L} \int_{\ln p_{O_2}^{''}}^{\ln p_{O_2}^{'}} t_{el} \sigma_{ion} d \ln p_{O_2} \tag{2}$$

Here σ_{ion} is the ionic conductivity, L the membrane thickness, and $p_{O_2}^{'}$ and $p_{O_2}^{''}$ are the oxygen partial pressures at the high and low oxygen partial pressure side, respectively. Other parameters have their usual significance. As $La_{1-x}Sr_xCoO_{3-\delta}$ shows predominant electronic conductivity, the electronic transference number, t_{el}, can be taken to be unity. Following data of the oxygen nonstoichiometry of phases $La_{1-x}Sr_xCoO_{3-\delta}$ [6], the ionic conductivity can be represented by the power law expression,

$$\sigma_{ion} = \sigma_{ion}^0 p_{O_2}{}^n \tag{3}$$

Substitution in Eq. 2 yields, after integration,

$$j_{O_2} = \frac{\sigma_{ion}^0 RT}{4^2 n F^2 L} \left[p_{O_2}^{'}{}^n - p_{O_2}^{''}{}^n \right] \tag{4}$$

To learn more about the kinetics of oxygen permeation, *i.e.* the relative role of surface exchange and bulk ionic diffusion in rate determining the transport of oxygen, measurements were conducted as a function of temperature, membrane thickness and oxygen partial pressure maintained at the oxygen-lean side of the membrane. Air was fed at the other side of the membrane in all cases.

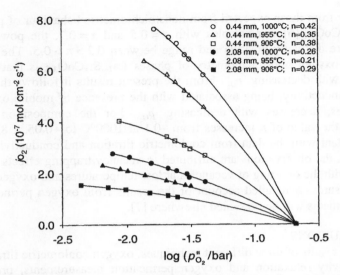

Fig. 4. Dependence of j_{O_2} on $\log(p''_{O_2})$ for La$_{0.5}$Sr$_{0.5}$CoO$_{3-\delta}$ measured using air as feed gas. The drawn lines represent the fit of the data to Eq. (4).

The results give clear evidence that, even for the thinnest specimens in the experimental range 0.5 - 2mm, the oxygen fluxes remain predominantly controlled by bulk oxygen diffusion across the membrane. The calculated characteristic membrane thickness L_c, below which oxygen transport is predominantly rate limited by surface exchange, is found to vary between $60 - 230\ \mu m$ at temperatures 750 - 1000°C. Fitting of the experimental data collected at constant p_{O_2}-gradients in the range of $\log(p'_{O_2} / p''_{O_2})$ from 0.7 to 2.0 yielded similar values for L_c.

Typical results from oxygen permeation measurements as a function of oxygen partial pressure maintained at the oxygen-lean side of the membrane are shown in Fig. 4. Similar curves were obtained for compositions with $x = 0.2$ and $x = 0.7$. Drawn lines in Fig. 4 correspond to the best fit of the experimental data to Eq. (4), using n and σ^0_{ion} as fitting parameters. All fitted curves were forced to pass the point where $p'_{O_2} = p''_{O_2}$ and, hence, the oxygen flux has reduced to zero. In derivation of Eq. 4, the assumption is made of non-interacting, fully ionised oxygen vacancies, all contributing to transport. However, the values obtained for the power index n differ strongly from those expected from nonstoichiometric behaviour of phases $La_{1-x}Sr_xCoO_{3-\delta}$ [6]. For samples with $x = 0.5$ and $x = 0.7$, the power indexes are distinctly positive and range between $0.2 > n > 0.5$. The trend for the oxygen nonstoichiometry of phases $La_{1-x}Sr_xCoO_{3-\delta}$ is that δ increases with decreasing p_{O_2}. From the present results it follows that the ionic conductivity, being associated with the presence of mobile oxygen vacancies, decreases with decreasing p_{O_2}. For the composition with $x = 0.2$, the value of n increases from -0.1 at 1000°C to +0.05 at 855°C. Concordant with the data from coulometric titration and conductivity relaxation, the observations are attributed to vacancy trapping effects associated with the ordering of vacancies at low temperatures and oxygen partial pressures. A detailed account of the results from oxygen permeation measurements will be presented elsewhere [7].

CONCLUSIONS

The results of three different techniques, oxygen coulometric titration, conductivity relaxation and oxygen permeation measurements, provide clear evidence that oxygen nonstoichiometry in phases $La_{1-x}Sr_xCoO_{3-\delta}$ is accommodated by vacancy ordering to a degree which depends on oxygen partial pressure, temperature and strontium-content.

REFERENCES

[1] H.J.M. Bouwmeester and A.J. Burggraaf, "Dense Ceramic Membranes for Oxygen Separation," pp. 481-553 in *CRC Handbook of Solid State Electrochemistry*. Edited by P.J. Gellings and H.J.M. Bouwmeester. CRC, Boca Raton, USA. (1997).

[2] M.H.R. Lankhorst, H.J.M. Bouwmeester and H. Verweij, "Use of the Rigid Band Formalism to interpret the Relationship between O Chemical Potential and Electron Concentration in $La_{1-x}Sr_xCoO_{3-\delta}$," *Phys. Rev. Lett.*, **77** [14] 2989-92 (1996).

[3] M.H.R. Lankhorst, H.J.M. Bouwmeester and H. Verweij, "High-temperature coulometric titration of $La_{1-x}Sr_xCoO_{3-\delta}$, "Evidence for the Effect of Electronic Band Structure on Nonstoichiometry Behavior," *J. Solid State Chem.*, **133** 555-67 (1997).

[4] S. Adler, S. Russek, J. Reimer, M. Fendorf, A. Stacey, Q. Huang, A. Santoro, J. Lynn, J. Baltisberger, U. Werner, "Local structure and oxide-ion motion in defective perovskites," *Solid State Ionics*, **68** 193 (1994).

[5]L.M. van der Haar,   M.W. den Otter,   M. Morskate,   H.J.M. Bouwmeester, H. Verweij, "Chemical Diffusion and Oxygen Surface Transfer of $La_{1-x}Sr_xCoO_{3-\delta}$ studied with Electrical Conductivity Relaxation," submitted to *J. Electrochem. Soc.*

[6]J. Mizusaki, Y. Mima, S. Yamauchi, K. Fueki, "Nonstoichiometry of the Perovskite-Type Oxides $La_{1-x}Sr_xCoO_{3-\delta}$," *J. Solid Sate Chem.* **80**, 102 (1989).

[7]L.M. van der Haar, H.J.M. Bouwmeester, H. Verweij "Oxygen permeation through $La_{1-x}Sr_xCoO_{3-\delta}$ membranes, " *manuscript in preparation.*

OXYGEN PERMEATION PROPERTIES OF PEROVSKITE-RELATED INTERGROWTH OXIDES IN THE Sr-Fe-Co-O SYSTEM

F. Prado, T. Armstrong, and A. Manthiram
Materials Science and Engineering Program, ETC 9.104
The University of Texas at Austin
Austin, TX 78712

ABSTRACT

$Sr_4(Fe,Co)_6O_{13+\delta}$ and the n = 1, 2, 3 members of the Ruddlesden-Popper (R-P) series $(La,Sr)_{n+1}(Fe,Co)_nO_{3n+1}$ crystallizing in perovskite-related intergrowth structures have been investigated for use as oxygen separation membranes. The oxygen permeation flux jO_2 of $Sr_4Fe_{6-y}Co_yO_{13+\delta}$ for y > 1.5 is due to a Fe-rich perovskite phase present as a secondary phase. The three R-P compounds differ in the oxygen permeation properties. While the $Sr_{3-x}La_xFe_{2-y}Co_yO_{7-\delta}$ (n = 2) and $LaSr_3Fe_{3-y}Co_yO_{10-\delta}$ (n = 3) phases show oxygen permeation flux values one order of magnitude lower than the $SrCo_{0.8}Fe_{0.2}O_{3-\delta}$ perovskite phase, the $La_{1-x}Sr_{1+x}Fe_{1-y}Co_yO_4$ (n = 1) phases show negligible oxygen permeability. The $Sr_{3-x}La_xFe_{2-y}Co_yO_{7-\delta}$ phases show good structural stability at high temperatures.

INTRODUCTION

The development of mixed conducting oxides for electrochemical applications at high temperature has drawn much attention in recent years [1-5]. Unlike pure electronic or oxide ion conductors, these compounds exhibit both electronic and oxide-ion conduction simultaneously, which makes them suitable for oxygen separation membranes. For example, they can be used as oxygen separation membranes in methane (CH_4) conversion reactors to obtain value added products such as formaldehyde and syn-gas [6] eliminating the necessity of electrodes and external circuitry of traditional ceramic oxygen pumps. However, as the membrane is exposed to severe working conditions (high temperature and highly reducing environment), the material has to exhibit good structural and chemical stability along with high oxide-ion conductivity. Although the perovskite oxides $(La,Sr)(Fe,Co)O_{3-\delta}$ with high Sr and Co contents exhibit high oxygen permeability rates [7] that make them attractive candidates for use as oxygen separation membranes, they tend to suffer from structural transformations at high temperature and are not chemically stable [8]. Alternatives to the perovskite oxides are the perovskite-related intergrowth oxides such as the $Sr_4Fe_6O_{13+\delta}$ phase and the n = 1, 2, and 3 members of the Ruddlesden-Popper series $(La,Sr)_{n+1}Fe_nO_{3n+1}$. In the R-P series, the value of n indicates the number of perovskite layers present between the rock salt layers. The crystal structures of these compounds are closely related to the perovskite structure. They have a layered structure in which SrO-FeO$_2$-SrO blocks (perovskite blocks) alternate with either $Fe_2O_{2.5}$ blocks as in $Sr_4Fe_6O_{13+\delta}$ (Figs. 1a) or SrO rock salt blocks as in the R-P phases $(La,Sr)_{n+1}Fe_nO_{3n+1}$ (Fig. 1b-1d). While the $Sr_4Fe_6O_{13+\delta}$ phase has drawn some attention [9,10] after the report of high oxygen permeability by Ma *et al* [11] for the composition $Sr_4Fe_4Co_2O_{13}$, little or no information is available on the R-P phases.

In this paper, we compare the oxygen permeation properties of the perovskite-related intergrowth oxides $Sr_4Fe_{6-y}Co_yO_{13+\delta}$ ($0.0 \leq y \leq 2.6$) and the R-P series members $La_{1-x}Sr_{1+x}Fe_{1-y}Co_yO_4$ (n = 1), $La_{3-x}Sr_xFe_{2-y}Co_yO_{7-\delta}$ (n = 2) and $LaSr_3Fe_{3-y}Co_yO_{10-\delta}$ (n = 3) with that of the perovskite phase $SrCo_{0.8}Fe_{0.2}O_{3-\delta}$. The synthesis, crystal structure, oxygen content, and transport properties are also presented.

EXPERIMENTAL

Perovskite-related intergrowth oxides with the general formula $(Sr,La)_{n+1}(Fe,Co)_nO_{3n+1-\delta}$ (n = 1, 2, 3 and ∞) and $Sr_4Fe_{6-y}Co_yO_{13+\delta}$ were synthesized by both standard solid state reactions as well as a sol-gel procedure at 1000-1400 °C [2,3,9,12]. The samples were characterized by X-ray diffraction, iodometric titration (oxygen content analysis), thermogravimetric analysis (TGA), and four-probe conductivity measurements. Oxygen permeation measurements were carried out with sintered, polished discs having a thickness of about 1.5 mm and densities > 90% of theoretical density. It involved an exposure of one side of the disc to air (pO$_2$') and the other side to a lower oxygen partial pressure (pO$_2$") controlled by flowing He and measurement of the oxygen flux with a gas chromatograph [9].

RESULTS AND DISCUSSION

$SrFe_{6-y}Co_yO_{13+\delta}$ System

To improve the Co solubility in the $Sr_4Fe_{6-y}Co_yO_{13+\delta}$ system, the samples were prepared by a sol-gel technique [9] for $0 \leq y \leq 3.0$. Figure 2 shows the X-ray powder diffraction patterns of the samples obtained by firing the gel at 1000 °C for 48 h in air. While the $Sr_4Fe_{6-y}Co_yO_{13}$ compositions are single phase with the intergrowth structure of $Sr_4Fe_6O_{13}$ (Fig. 1a) for $y \leq 1.5$, they consist of three phases for $y \geq 1.8$: a perovskite-related intergrowth oxide $Sr_4Fe_{6-y}Co_yO_{13+\delta}$ with $y < 1.5$, a perovskite phase $SrFe_{1-y}Co_yO_{3-\delta}$, and a spinel phase $Co_{3-y}Fe_yO_4$. The

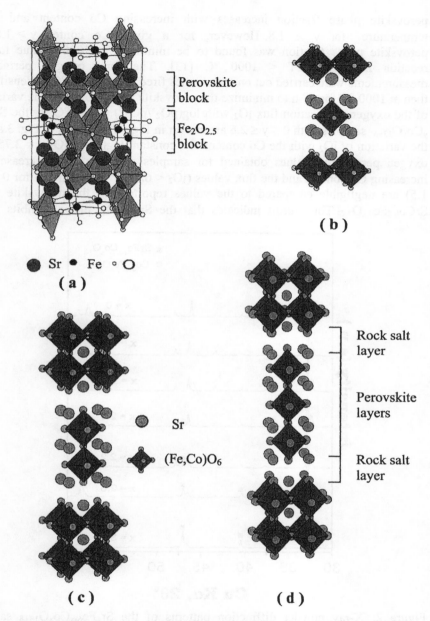

Figure 1: Crystal Structures of (a) Sr₄Fe₆O₁₃, (b) Sr₂FeO₄, (c) Sr₃Fe₂O₇, and (d) Sr₄Fe₃O₁₀.

perovskite phase fraction increases with increasing Co content and firing temperature for $y \geq 1.8$. However, for a given Co content $y > 1.8$, the perovskite phase fraction was found to be minimum at 1000 °C due to slow reaction kinetics at $T < 1000$ °C [13]. Therefore, oxygen permeation measurements were carried out on membranes fired first at 1150°C to densify and then at 1000 °C for 48 h to minimize the perovskite phase content. The variations of the oxygen permeation flux jO_2 with $\log(pO_2'/pO_2")$ at 900 °C for the $Sr_4Fe_{6-y}Co_yO_{13+\delta}$ samples with $0 \leq y \leq 2.6$ are shown in Fig. 3. The inset in Fig. 3 shows the variation of jO_2 with the Co content at a constant $\log(pO_2'/pO_2") = 2.75$. The oxygen permeation values obtained for samples with $y \leq 1.5$ decrease with increasing Co content and the flux values ($jO_2 < 6 \times 10^{-10}$ mol cm^{-2} s^{-1} for $0 \leq y \leq 1.5$) are negligible compared to the values reported for the perovskite phase $SrCo_{0.8}Fe_{0.2}O_{3-\delta}$. This result indicates that the $Sr_4Fe_6O_{13}$ phase exhibits poor

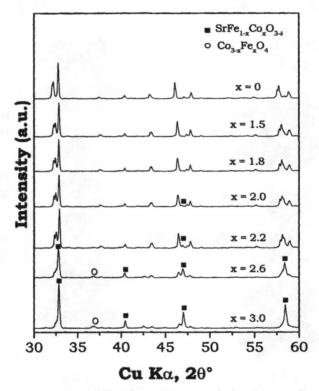

Figure 2: X-ray powder diffraction patterns of the $Sr_4Fe_{6-x}Co_xO_{13+\delta}$ samples obtained by firing the gel at 1000 °C for 48 h in air. The unmarked reflections refer to those from the intergrowth phase.

oxide-ion conduction even after doping with Co. The $Sr_4Fe_{6-y}Co_yO_{13+\delta}$ phases with $0 \leq y \leq 1.5$ are stable in both air and N_2 atm and they do not lose much oxygen, which results in a negligible oxygen concentration gradient across the membrane and low jO_2 values. On the other hand, jO_2 increases with increasing Co content for $y \geq 1.5$, but the improvement in jO_2 takes place with an increase in the perovskite phase fraction. Therefore, we conclude that the higher oxygen permeability of the samples with $y \geq 1.5$ is due to an increasing amount of the perovskite phase, which is known to show oxygen flux values two orders of magnitude higher than the intergrowth oxide $Sr_4Fe_{6-y}Co_yO_{13+\delta}$ [9].

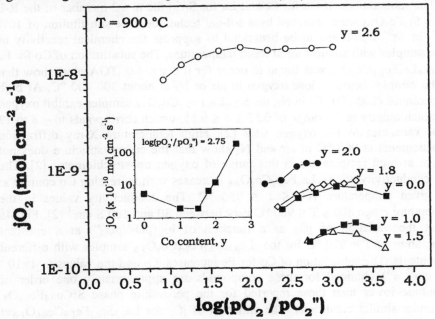

Figure 3: Oxygen permeation flux as a function of log (pO_2'/pO_2'') for $Sr_4Fe_{6-y}Co_yO_{13+\delta}$ with $0 \leq y \leq 2.6$. The inset displays jO_2 as a function of y at constant log (pO_2'/pO_2'') = 2.75.

La$_{1-x}$Sr$_{1+x}$BO$_4$ (B = Fe, and Co) System

Dense pellets (> 90%) annealed in air for oxygen permeation measurements could be obtained only for x = 0 and 0.2 in both the cases of $La_{1-x}Sr_{1+x}FeO_4$ and $La_{1-x}Sr_{1+x}CoO_4$. At room temperature, the oxygen content of these samples is ~ 4.0. TGA plots show the oxygen content does not change with temperature for Fe-containing samples or changes slightly (~ 2%) for Co-containing samples up to

900 °C in either flowing air or N_2. The jO_2 values increase with the substitution of Sr for La and Co for Fe. However, the jO_2 values obtained for both the systems are between 1×10^{-10} and 4×10^{-9} mol cm^{-2} s^{-1} for $1.0 \leq$ log (pO$_2$'/pO$_2$") ≤ 3.5 [12]. These values are 3 to 4 orders of magnitude lower than that reported for the $La_{1-x}Sr_xCo_{1-y}Fe_yO_{3-\delta}$ perovskite phases [7]. The low oxygen vacancy concentration revealed from TGA experiments seems to lead to low oxygen permeation flux values at high temperatures in these systems.

$Sr_{3-x}La_xFe_{2-y}Co_yO_{7-\delta}$ System

The substitutions of Co for Fe and La for Sr in the n = 2 member of the R-P series $Sr_3Fe_2O_{7-\delta}$ were achieved by a sol-gel technique. The substitution of 10% La^{3+} for Sr^{2+} was found to be beneficial to suppress the chemical reactivity of these samples with ambient air at room temperature. The substitution of Co for Fe in $Sr_{3-x}La_xFe_{2-y}Co_yO_{7-\delta}$ was found to occur for $0 \leq y \leq 1.0$. TGA plots show that all the samples begin to lose oxygen in air or N_2 at about 300-400 °C. At high temperatures (800-900 °C) in N_2, the $Sr_{3-x}La_xFe_{2-y}Co_yO_{7-\delta}$ samples exhibit oxygen nonstoichiometry in the range of $0.55 \leq \delta \leq 0.85$, which corresponds to ~ 8 to 12 % of vacancies in the oxygen sites [2]. High temperature X-ray diffraction measurements at 900 °C in air and N_2 show that the crystal structure does not change at high temperature in this range of oxygen nonstoichiometry [2]. The total conductivity of $Sr_{3-x}La_xFe_{2-y}Co_yO_{7-\delta}$ increases with increasing Co content at any given temperature $100 \leq T \leq 950$ °C. The conductivity values in the temperature range $800 \leq T \leq 900$ °C vary between 30 and 100 S cm^{-1} [2]. Fig. 4a shows the variation of jO_2 as a function of log(pO$_2$'/pO$_2$") at a constant temperature of T = 900 °C for the $La_{0.3}Sr_{2.7}Fe_{2-y}Co_yO_{7-\delta}$ samples with different Co contents. The substitution of Co for Fe increases jO_2 and the values (~ 1×10^{-7} mol cm^{-2} s^{-1}) obtained for these compounds are approximately one order of magnitude lower than those reported for the perovskite phase $SrCo_{0.8}Fe_{0.2}O_{3-\delta}$ [7] under similar conditions. The variation of jO_2 for $La_{0.3}Sr_{2.7}Fe_{1.4}Co_{0.6}O_{7-\delta}$ at constant temperature (800 and 900 °C) with the inverse of sample thickness (L) at constant pO$_2$ differences (log(pO$_2$'/pO$_2$") = 1.0 and 2.2) across the membranes shows a linear dependence with a good extrapolation to the origin. This behavior is in agreement with the Wagner equation that predicts a linear dependence of jO_2 with L^{-1} when jO_2 is bulk limited [2]. Therefore, the oxygen flux of $La_{0.3}Sr_{2.7}Fe_{1.4}Co_{0.6}O_{7-\delta}$ is bulk limited for $L \geq 0.5$ mm and higher jO_2 values can be expected by reducing L until the surface exchange limited regime is reached. Assuming $Sr_3Fe_2O_{7-\delta}$ and $La_{0.3}Sr_{2.7}Fe_{2-y}Co_yO_{7-\delta}$ with y = 0 and 1 are also bulk limited, the oxide-ion conductivity σ_i at 900 °C has been estimated. Considering that the electronic conductivity $\sigma_e \gg \sigma_i$ for both of the compositions and perovskite phases and assuming σ_i slightly changes in the pO_2 range of our measurements, the Wagner's equation can be simplified as

$$jO_2 = \frac{RT}{16F^2L}\overline{\sigma}_i \ln(pO_2'/pO_2'')$$ (1)

where R is the gas constant, F is the Faraday constant, T is the temperature, and L is the thickness of the membrane. Using the relation (1) and the experimental data in Fig. 4a, σ_i values ranging between 5×10^{-3} S cm^{-1} (transference number $t_i = 1.6 \times 10^{-4}$) for La$_{0.3}$Sr$_{2.7}$Fe$_2$O$_{7-\delta}$ and 4.5×10^{-2} S cm^{-1} ($t_i = 5.8 \times 10^{-4}$) for La$_{0.3}$Sr$_{2.7}$FeCoO$_{7-\delta}$ were obtained [2].

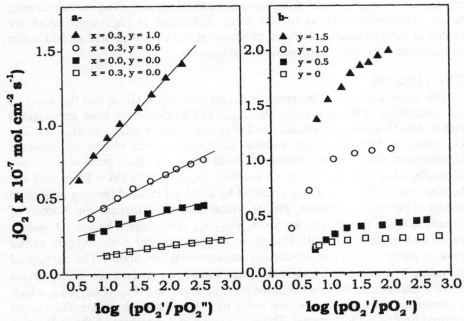

Figure 4: Variations of jO₂ with log (pO₂'/pO₂") for (a) Sr₃₋ₓLaₓFe₂₋yCoyO₇₋δ, and (b) LaSr₃Fe₃₋yCoyO₁₀₋δ. Measurements were conducted at 900 °C.

LaSr$_3$Fe$_{3-y}$Co$_y$O$_{10-\delta}$ System

The LaSr$_3$Fe$_{3-y}$Co$_y$O$_{10-\delta}$ phases belonging to the n = 3 member of the R-P series were prepared by solid-state reactions. X-ray diffraction analysis shows that single-phase products are formed for $0.0 \leq y \leq 1.5$. These samples are tetragonal (space group I4/mmm) and the lattice parameters decrease with increasing Co content [3]. TGA analysis shows that the samples lose considerable amount of oxygen at temperatures T > 350 °C in both air and N_2 atm. At high temperatures (800 - 900 °C), the oxygen nonstoichiometry δ is ~ 0.3-0.45 in air and ~ 0.75 in N_2 atm, which corresponds to 7.5% vacancies in the oxygen sites. The total conductivity increases with increasing Co content and the values obtained at high temperature are between 50 and 100 S cm^{-1} [12]. The variations of oxygen permeation flux at 900 °C with log pO_2 gradient across the sample are shown in Fig. 4b. The data reveal that the oxygen permeation flux increases with increasing Co content. The jO_2 values obtained for $LaSr_3Fe_{3-y}Co_yO_{10-\delta}$ are ~ 1×10^{-7} mol cm^{-2} s^{-1}, which are one order of magnitude lower than those reported for the perovskite phase $SrFe_{0.2}Co_{0.8}O_{3-\delta}$ and are similar to the values obtained for the $La_{0.3}Sr_{2.7}Fe_{2-y}Co_yO_{7-\delta}$ samples. Although no evidence of crystal structure transformation was detected from oxygen permeation and TGA measurements, further experiments such as *in situ* X-ray diffraction at high temperature are needed to fully evaluate the crystal structure of the Co-substituted phases under the conditions of oxygen permeation measurements.

CONCLUSIONS

The perovskite-related intergrowth oxides $Sr_4Fe_{6-y}Co_yO_{13+\delta}$, and the n = 1, 2 and 3 members of the R-P series $(Sr,La)_{n+1}(Fe,Co)_nO_{3n+1}$ have been investigated with an aim to develop structurally and chemically stable mixed conductors. The jO_2 values at 900 °C of the various intergrowth oxide phases for selected compositions have been compared with that of the perovskite phase $SrCo_{0.8}Fe_{0.2}O_{3-\delta}$ (Fig. 5). The n = 1 member $La_{0.8}Sr_{1.2}MO_4$ (M = Fe or Co) and the composition $Sr_4Fe_4Co_2O_{13+\delta}$ prepared by a sol-gel method (having negligible amount of perovskite secondary phase) exhibit flux values that are 2 to 3 orders of magnitude lower than that of the $SrCo_{0.8}Fe_{0.2}O_{3-\delta}$ perovskite. The low jO_2 values are related to the negligible change in oxygen content with oxygen partial pressure under the oxygen permeation measurement conditions. The compound $Sr_4Fe_4Co_2O_{13+\delta}$ prepared by solid-state reaction shows an increase in jO_2 values due to an increase in the amount of the perovskite phase $SrFe_{0.75}Co_{0.25}O_{3-\delta}$, which is known to show jO_2 values one order of magnitude higher than the sol-gel prepared $Sr_4Fe_4Co_2O_{13}$ material. The n = 2 and n = 3 members of the R-P series show jO_2 values comparable to that of some perovskite phases like $SrFe_{0.8}Co_{0.2}O_{3-\delta}$ and $La_{0.4}Sr_{0.6}Co_{0.8}Fe_{0.2}O_{3-\delta}$ [2], but approximately one order of magnitude lower than that of the perovskite phase $SrCo_{0.8}Fe_{0.2}O_{3-\delta}$. The n = 2 and n = 3 compounds show both higher oxygen vacancy concentration and higher total conductivity (σ > 30 S cm^{-1}) than the n = 1 compounds. Also, the n = 2 and n = 3 compounds are stable under permeation measurement conditions. The better stability of the intergrowth oxides despite the lower oxygen flux may make them attractive for electrochemical applications.

Materials for Electrochemical Energy Conversion and Storage

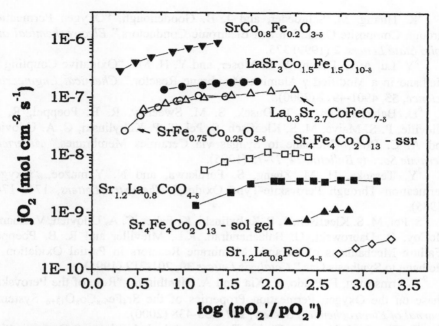

Figure 5: Comparison of the variation of jO$_2$ with log(pO$_2$'/pO$_2$") at 900 °C of selected intergrowth oxides with that of the perovskite phases SrFe$_{1-y}$Co$_y$O$_{3-\delta}$ with y = 0.25 and 0.8. All the samples are 1.5 mm thick.

ACKNOWLEDGEMENT

Acknowledgment is made to the Welch Foundation (Grant No. F-1254) for the support of this research. One of the authors (F. P.) thanks CONICET, Argentina, for a postdoctoral fellowship.

REFERENCES

[1] Y. Tsuruta, T. Todaka, H. Nisiguchi, T. Ishihara, and Y. Takita, "Mixed Electronic-Oxide Ionic Conductor of Fe-Doped La(Sr)GaO$_3$ Perovskite Oxide for Oxygen Permeating Membrane," *Electrochemical and Solid State Letters*, **4**, E13-E15 (2001).

[2] F. Prado, T. Armstrong, A. Caneiro, and A. Manthiram, "Structural Stability and Oxygen Permeation Properties of Sr$_{3-x}$La$_x$Fe$_{2-y}$Co$_y$O$_{7-\delta}$ (0.0 ≤ x ≤ 0.3 and 0.0 ≤ y ≤ 1.0)," *Journal of Electrochemical Society*, **148**, J7-J14 (2001).

[3] T. Armstrong, F. Prado, and A. Manthiram, "Synthesis, Crystal Chemistry, and Oxygen Permeation Properties of LaSr$_3$Fe$_{3-x}$Co$_x$O$_{10}$ (0 ≤ x ≤ 1.5)," *Solid State Ionics* **140**, 89-96 (2001).

[4]K. Huang, M. Schroeder, and J. B. Goodenough, "Oxygen Permeation through Composite Oxide-Ion and Electronic Conductors," *Electrochemical and Solid State Letters*, **2** (1999) 375.

[5]Y. Lu, A. G. Dixon, W. R. Moser, and Y. H. Ma, "Oxidative Coupling of Methane in a Modified γ-Alumina Membrane Reactor," *Chemical Engineering Science*, **55**, 4901-4912 (2000).

[6]U. Balachandran, J. T. Dusek, S. M. Sweeney, R. B. Poeppel, R. L. Mieville, P. S. Maiya, M. S. Kleefisch, S. Pei, T. P. Kobylinski, C. A. Udovich and A. C. Bose, "Methane to Syngas via Ceramics Membranes," *American Ceramic Society Bulletin*, **74**, 71-75 (1995).

[7]Y. Teraoka, H. M. Zhang, S. Furukawa, and N. Yamazoe, "Oxygen Permeation Through Perovskite-Type Oxides," *Chemistry Letters*, 1743-1746 (1985).

[8]S. Pei, M. S. Kleefisch, T. P. Kobylinski, K. Faber, C. A. Udovich, V. Zhang-McCoy, B. Dabrowski, U. Balachandran, R.L. Mieville, and R. B. Poeppel, "Failure Mechanisms of Ceramic Membrane Reactors in Partial Oxidation of Methane to Synthesis Gas," *Catalysis Letters* **30**, 201-212 (1995).

[9]T. Armstrong, F. Prado, Y. Xia ,and A. Manthiram, "Role of the Perovskite Phase on the Oxygen Permeation Properties of the $Sr_4Fe_{6-x}Co_xO_{13+\delta}$ System," *Journal of Electrochemical Society*, **147**, 435-438 (2000).

[10]S. Kim, Y. L Yang, R. Christoffersen, and A. J. Jacobson, " Determination of Oxygen Permeation Kinetics in a Ceramic Membrane Based on the Composition $SrFeCo_{0.5}O_{3.25-\delta}$," *Solid State Ionics*, **109**, 187-196 (1998).

[11]B. Ma, U. Balachandran, J.-H. Park, C. U. Segre, "Electrical Transport Properties and Defect Structure of $SrFeCo_{0.5}O_x$," *Journal of Electrochemical Society* **143**, 1736- (1996).

[12]T. Armstrong, "Oxygen Permeation Properties of Perovskite-Related Intergrowth Oxides Exhibiting Mixed Ionic-Electronic Conduction," *Ph. D. Dissertation*, University of Texas at Austin, Austin, TX, 2000.

[13]Y. Xia, T. Armstrong, F. Prado, and A. Manthiram, "Sol-gel Synthesis, Phase Relationships, and Oxygen Permeation Properties of $Sr_4Fe_{6-x}Co_xO_{13+\delta}$ ($0 \leq x \leq 3$)," *Solid State Ionics* **130**, 81-90 (2000).

Fe DOPED LaGaO₃ BASED PEROVSKITE OXIDE AS AN OXYGEN SEPARATING MEMBRANE FOR CH₄ PARTIAL OXIDATION

Tatsumi Ishihara, Yuko Tsuruta, Hiroyasu Nishiguchi and Yusaku Takita
Department of Applied Chemistry, Faculty of Engineering, Oita University, Dannoharu 700, Oita 870-1192, Japan

ABSTRACT

Oxygen permeating property of Fe doped $LaGaO_3$ based perovskite oxide was investigated in this study. Fe doped $La_{1-x}Sr_xGaO_3$ exhibits high total conductivity (~ 1 S/cm) and high oxygen permeation rate. In particular, the highest total conductivity and oxygen permeating rate were attained at $La_{0.7}Sr_{0.3}Ga_{0.6}Fe_{0.4}O_3$ (LSGF). At this composition, oxygen permeation rate from air to He was as high as $102 \mu mol/mincm^{-2}$ ($2.5cc/min \cdot cm^{-2}$) at 1273 K and 0.3 mm membrane thickness. Application of LSGF membrane for CH_4 partial oxidation was further studied. Since the P_{O2} difference became large, oxygen permeation rate drastically increased and it attained at 492 $\mu mol/min\ cm^2$ (12 cc/min.cm²) at 1273 k, 0.5mm in thickness. Since no significant change was observed on the XRD pattern of LSGF after CH_4 partial oxidation, it seems like that stability of LSGF against reduction is high enough for the partial oxidation of CH_4 .

INTRODUCTION

Methane (CH_4) is the major component in a natural gas which is an abundant natural resource. Therefore, conversion of CH_4 into a useful compound is an important subject at present. From the view point of the useful utilization of CH_4, partial oxidation of CH_4 is attracting much attention, since the reaction gives a synthesis gas at $CO:H_2=1:2$ which is suitable for the synthesis of methanol or hydrocarbons. Pure oxygen gas is an essential reactant for this reaction and as far as the cost is concerned, separation of air into O_2 and N_2 by a simple method should be considered. Separation of air into O_2 and N_2 by a mixed electronic and oxide ionic conducting ceramic membrane is an ideal method for obtaining pure oxygen because of its simple structure and low energy consumption [1]. It is well-known that $La(Sr)Fe(Co)O_3$ (LSFC) perovskite oxides exhibit a superior mixed conductivity and consequently, high oxygen permeation rate is attained on these perovskite oxides[2]. However, LSFC is easily reduced in CH_4 atmosphere and it is reported that failure of membrane due to reduction sometimes occurs when LSFC is used as the oxygen permeating membrane for CH_4 partial oxidation[3]. On the other hand, it is reported that $SrFeCo_{0.5}O_3$ is stable against reduction and it can be used for the oxygen permeating membrane for the partial oxidation of methane[4,5]. However, further improvement in oxygen permeating rate is required for the application of the mixed electronic-oxide ionic conductor to the oxygen generator for CH_4 partial oxidation.

In our previous study, it was found that Ni or Fe doped $La(Sr)GaO_3$ was a mixed electronic hole and oxide ionic conductor and the material was stable over a wide range of oxygen partial pressure (P_{O2})[6,7]. However, for the application of CH_4 partial oxidation, stability of Ni doped $LaGaO_3$ based oxides in reducing atmosphere is not high enough. From the stability point of view, Fe doped $La(Sr)GaO_3$ seems to be a better candidate as the mixed electronic and oxide ionic conductor for CH_4 partial oxidation. In this study, therefore, mixed electronic and oxide ionic conductivity in Fe doped $La(Sr)GaO_3$ was investigated.

EXPERIMENTAL
Preparation of Fe doped $LaGaO_3$ was performed by a conventional solid state reaction using La_2O_3 (99.99%), $SrCO_3$ (reagent grate), Ga_2O_3 (99.99%), and Fe_2O_3 (99.5%). Metal oxides at desired composition was prepared in an Al_2O_3 mortal and pestle and subsequently, mixture was calcined at 1273K for 6h. The   powder

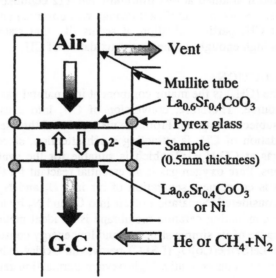

Fig.1 Experimental set-up of membrane reactor used for CH_4 partial oxidation

after calcination was mixed again and pressed into a disk (20mm in diameter) followed by iso-static pressing at 2700 kg/cm^2 for 20min and obtained disks were sintered at 1773K for 6h. Finally, disks were ground to 0.5mm in thickness with diamond wheel. Each disks were painted on both surfaces with La$_{0.6}$Sr$_{0.4}$CoO$_3$ (LSC) slurry at 10mm in diameter in order to improve the surface activity for oxygen dissociation. LSC was prepared by calcination of the mixture of La(NO$_3$)$_3$, Sr(NO$_3$)$_2$, and (CH$_3$COO)$_2$Co at 1273K for 6h. Oxygen gas concentration cell of air-He was used for oxygen permeating measurement. Permeating oxygen from air to He was analyzed by using a gas chromatograph. For the partial oxidation of CH$_4$, Ni and LSC were coated on the membrane as partial oxidation and oxygen dissociation catalyst, respectively.   Pyrex glass ring was used for sealing gas as shown in Fig.1. Electrical conductivity was measured with DC four-probe method in the gas flow cell. Transport number of oxide ion was estimated by the ratio of the measured electromotive force in H$_2$-O$_2$ cell to that estimated by Nernst equation.

RESULTS AND DISCUSSION
Oxygen Permeation from Air to Ar
Figure 2 shows the Arrhenius plots of electrical conductivity of La$_{1-x}$Sr$_x$Ga$_{0.6}$Fe$_{0.4}$O$_3$ in P$_{O2}$=10^{-5}atm.   It is obvious that all specimens exhibit a high total conductivity like 1~10 S/cm.   The total conductivity increased with

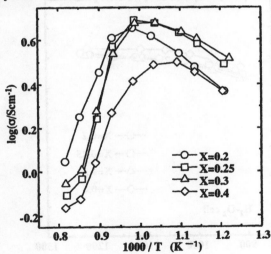

Fig.2 Arrhenius plots of electrical conductivity of La$_{1-x}$Sr$_x$Ga$_{0.6}$Fe$_{0.4}$O$_3$ in P$_{O2}$=10^{-5}atm.

increasing the amount of Sr and attained the maximum at X=0.3. On the other hand, conductivity of all specimens exhibited metal like temperature dependence at temperature higher than 973K. Namely conductivity decreased despite increasing temperature. In any case, it can be said that $La_{1-x}Sr_xGa_{0.6}Fe_{0.4}O_3$ exhibits a high total conductivity. In a range of P_{O2} higher than $P_{O2}=10^{-5}$atm, the electrical conductivity decreased with increasing oxygen partial pressure and P_{O2} dependence of electrical conductivity is almost $P_{O2}^{1/4}$. Consequently, it is clear that the main electronic charge carrier in $La_{1-x}Sr_xGa_{0.6}Fe_{0.4}O_3$ is electronic hole, which could be assigned to the formation of Fe^{4+}. On the other hand, electronic conductivity of $La_{1-x}Sr_xGa_{0.6}Fe_{0.4}O_3$ was almost independent of oxygen partial pressure in a range of $P_{O2}=10^{-5} \sim 10^{-15}$atm. Therefore, in this P_{O2} range, it seems likely that the amount of Fe^{4+} ($[Fe_{Ga}']$)is balanced with that of interstitial excess oxygen ($[O_i^{\cdot\cdot}]$).

Temperature dependence of transport number of oxide ion in $La_{1-x}Sr_xGa_{0.6}Fe_{0.4}O_3$, which was estimated with electromotive force in H_2-O_2 cell, is shown in Figure 3. Obviously, electromotive forces in H_2-O_2 cell using $La_{1-x}Sr_xGa_{0.6}Fe_{0.4}O_3$ were almost half of those estimated by Nernst equation, and furthermore it changed with P_{O2} difference. Therefore, it can be said that Fe doped $LaGaO_3$ exhibits a high oxide ionic conduction in simultaneous with electric hole. Since the largest permeation rate of oxygen is theoretically achieved at the

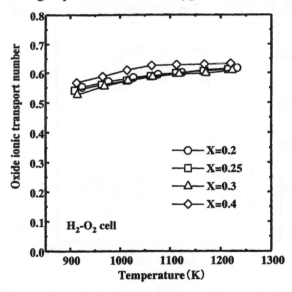

Fig.3 Temperature dependence of transport number of oxide ion in $La_{1-x}Sr_xGa_{0.6}Fe_{0.4}O_3$.

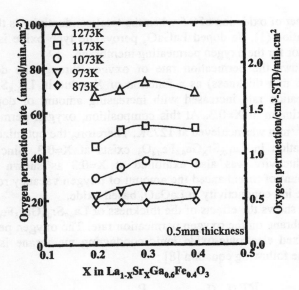

Fig.4 Permeation rate of oxygen through Fe doped $LaGaO_3$ membrane (0.5 mm thickness) as a function of X value in $La_{1-x}Sr_xGa_{0.6}Fe_{0.4}O_3$.

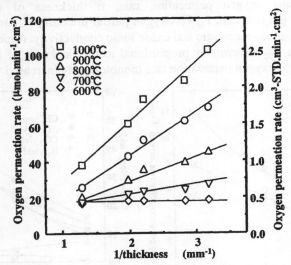

Fig. 5 Effects of the thickness of $La_{0.7}Sr_{0.3}Ga_{0.6}Fe_{0.4}O_3$ membrane on the oxygen permeation rate.

transport number of oxide ion of 0.5 when the total conductivity is the same as shown in equation (1), Fe doped $LaGaO_3$ perovskite type oxide is suitable as a mixed conductor for the oxygen permeating membrane.

Figure 4 shows the permeation rate of oxygen through Fe doped $LaGaO_3$ membrane (0.5 mm thickness) as a function of X value in $La_{1-x}Sr_xGa_{0.6}Fe_{0.4}O_3$. Oxygen permeation rate increased with increasing amount of doped Sr and it attained a maximum at X=0.3. At this composition, oxygen permeation rate of 1.8cc-STD/cm^2·min was achieved at 1273K. Therefore, the optimized X value for oxygen permeation in $La_{1-x}Sr_xGa_{0.6}Fe_{0.4}O_3$ exists at X=0.3. Since the highest electrical conductivity was also exhibited at X=0.3 as shown in Figure 2, increasing amount of Sr enhanced the amount of oxygen vacancy resulting in the improved oxide ion conductivity in $LaGaO_3$ based oxide.

Figure 5 shows the effects of the thickness of $La_{0.7}Sr_{0.3}Ga_{0.6}Fe_{0.4}O_3$ (denoted as LSGF) membrane on the oxygen permeation rate. The oxygen permeation rate through the mixed electronic-oxide ionic conducting membrane is theoretically expressed by the following equation [8] ;

$$J_{O2} = \frac{RT\sigma_e\sigma_i}{16F^2(\sigma_e+\sigma_i)t} \ln\frac{P_h}{P_l} \tag{1}$$

where, J_{O2} means oxygen permeation rate, t: thickness of membrane, T: temperature, R: gas constant, P_h, P_l: oxygen partial pressure, F: Faraday constant, and σ_e and σ_i mean electronic and oxide ionic conductivity, respectively. Based on this equation, J_{O2} is inversely proportional to the thickness of membrane. As shown in Figure5, oxygen permeation rate monotonously increased with

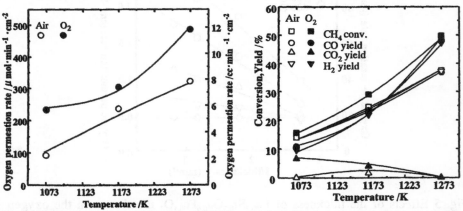

Fig.6 Temperature dependence of CH$_4$ partial oxidation in membrane reactor using LSGF for oxygen permeating membrane.

Materials for Electrochemical Energy Conversion and Storage

decreasing thickness of the membrane and the oxygen permeating rate at 1173K attained to a valve of 102 μmol/cm^{2}min(2.5cc-STD/cm^2·min) when the thickness of the LSGF membrane was 0.3 mm. Therefore, rate determining step for oxygen permeation through LSGF membrane seems to be bulk diffusion process and consequently, it is expected further higher oxygen permeation rate is obtained by decreasing thickness of membrane.

Application of LSGF for oxygen permeating membrane for CH$_4$ partial oxidation

Application of LSFC for oxygen permeating membrane for CH$_4$ partial oxidation was further studied. Figure 6 shows the temperature dependence of oxygen permeation rate, CH$_4$ conversion and yield of CO and H$_2$. It is clear that the oxygen permeation rate increased drastically by changing Ar to CH$_4$ for the permeating site gas as is predicted from the equation (1). The permeation rate at 1273 K was attained a value of ca.327μmol/min cm^2(8 cc/min cm^2). On the other hand, CH$_4$ conversion of 40 % and yield of CO and H$_2$ at the same value to CH$_4$ conversion were also obtained at and CO$_2$ formation was negligibly small, it is obvious that CH$_4$ partia1273 K. Since the yield of CO and H$_2$ were almost the same as that of CH$_4$ conversion l oxidation dominantly proceeds in the membrane reactor using LSGF as the oxygen permeating membrane.

Effects of oxygen partial pressure in oxidant on the permeation rate through LSGF membrane under CH$_4$ partial oxidation were further studied and results were compared in Fig.6. In is reasonable that oxygen permeation rate further increased by changing from air to oxygen at all temperatures examined. This suggests that the rate determining step of oxygen permeation under CH$_4$ partial oxidation is still bulk diffusion process. The amount of oxygen permeation was as high as 492 (12 cc/min cm^2) and 245 μmol/min cm^2 (6 cc/min cm^2) at 1273 and 1073 K, respectively. CH$_4$ conversion was also improved by changing air to oxygen. However, formation of CO$_2$ was observed and it became significant with decreasing reaction temperature when oxygen was used as oxidant. This suggests that the surface activity of the used Ni catalyst for the partial oxidation of CH$_4$ was not high enough, and the amount of permeating oxygen became excess when oxygen was used. Therefore, if the activity of Ni can be improved by improving dispersion, further higher yield of CO and H$_2$ may be obtained. In any way, even at 873 K, the rate determining step for oxygen permeating seem to be the bulk diffusion and so, it is expected that the oxygen permeation rate was further improved by decreasing the thickness of the LSGF membrane.

Figure 7 shows the oxygen permeation rate and the product yield in CH$_4$ partial oxidation as a function of membrane thickness at 1273 K. It is clear that the oxygen permeation rate monotonously increased with decreasing the thickness of the LSGF membrane. These results also suggested that the rate determining steps for CH$_4$ partial oxidation is bulk diffusion step even at 0.3 mm thickness membrane. The permeation rate from air to CH_4 was attained 492 μmol/min cm2(12 cc/min cm^2) at 1273 K as shown in Fig.7. In accordance with the increase in the amount of permeating oxygen, CH_4 conversion as well as yield of CO and H_2 also increased monotonously. Since the amount of carbon was almost balanced before and after the reactor, no carbon formation can be recognized on Ni catalyst by eye view, and H_2/CO ratio was always close to 2, it seems like that no carbon was deposited under CH_4 partial oxidation with LSGF membrane reactor and CH_4 partial oxidation into CO/H_2 gas was dominantly occurred. Therefore, CH_4 partial oxidation process combining with the oxygen separating membrane using mixed oxide ion conductor is highly attractive for the conversion of CH_4 into liquid fuels, since the size of reactor is small and process is also simple. It was obvious that no significant difference was observed on the XRD patterns of the LSGF film before and after the reaction. Therefore, LSGF membrane has enough stability against reduction.

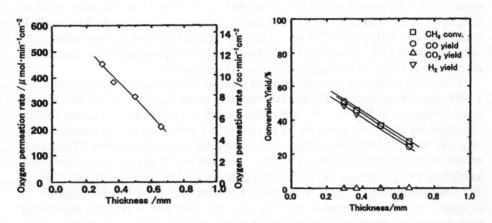

Fig. 7 Oxygen permeation rate and the product yield in CH_4 partial oxidation as a function of membrane thickness at 1273 K.

CONCLUSION

Although mixed oxide ion conductivity in $LaGaO_3$ based oxide have not been studied extensively, it was found that Fe doped $LaGaO_3$ based oxide exhibited high mixed electronic hole and oxide ion conductivity. Therefore, high oxygen permeating rate was achieved on LSGF membrane. Although $La_{0.6}Sr_{0.4}CoO_3$ and Ni are required for oxygen dissociation and CH_4 partial oxidation catalysts, rate determining step for CH_4 partial oxidation is bulk diffusion steps of oxygen through membrane at 873 K and 0.3mm thickness. Therefore, the oxygen permeation rate was further improved by decreasing the thickness of the LSGF

Materials for Electrochemical Energy Conversion and Storage

membrane.   Since $LaGaO_3$ based oxide have a high stability against reduction, Fe doped $LaGaO_3$, $La_{0.7}Sr_{0.3}Ga_{0.6}Fe_{0.4}O_3$, is highly attractive as the oxygen permeating membrane.

ACKNOWLEDGEMENT

The authors acknowledged the financial support from Grant-in-Aid for Science Promotion from Ministry of Education, Culture, Sports, Science and Technology of Japan.

REFERENCES

[1]S.Pei, M.S.Kleefisch, T.P.Kobylinski, J.Faber, C.A.Udovich, V.Zhang-McCoy, B.Dabroaski, U.Balachandran, R.L.Mieville and R.B.Poeppel, Catalysis Lett., 30, 201 (1995).

[2]C.-Y.Tsai, A.G.Dixon, Y.H.Ma, W.R.Moser and M.R.Pascucci, J. Am. Ceram. Soc., 81, 1437 (1998).

[3]Y.Teraoka, T.Nobunaga., and N.Yamazoe, Chem.Lett., 1 (1990).

[4]B.Ma, U.Balachandran, J.-H.Park, and C.U.Segre, Solid State Ionics, 83, 65 (1996).

[5]U.Balachandran, J.T.Dusek, S.M.Sweeney, R.B.Poeppel, R.L.Mieville, P.S.Maiya, M.S.Kleefish, S.Pei, T.P.Kobylinski, and C.AA.Udovich, Am.Cerm.Soc.Bull., 74, 71 (1995).

[6]H.Arikawa, T.Yamada, T.Ishihara, H.Nishiguchi, and Y.Takita, Chem Lett., 1257 (1999)

[7]T.Ishihara, T.Yamada, H.Arikawa, H.Nishiguchi and Y.Takita, Solid State Ionics,135, 631 (2000).

[8]N.Itoh, T.Kato, K.Uchida and K.Haraya, Journal of Membrane Science, 92, 239 (1994).

SYNTHESIS AND OXYGEN PERMEATION PROPERTIES OF $Sr_{2.7}La_{0.3}Fe_{2-y}M_yO_{7-\delta}$ (M = Mn, Co and Ni)

F. Prado and A. Manthiram
Materials Science and Engineering Program, ETC 9.104
The University of Texas at Austin
Austin, TX 78712

ABSTRACT

Perovskite-related intergrowth oxides $Sr_{2.7}La_{0.3}Fe_{2-y}M_yO_{7-\delta}$ (M = Mn, Co and Ni) have been synthesized by a sol-gel method for $0 \leq y \leq 1.0$. The samples have been characterized by X-ray powder diffraction, thermogravimetric analysis, and transport measurements. The substitution of Mn, Co or Ni for Fe does not change the tetragonal symmetry of the crystal structure at room temperature. Both the electrical conductivity and oxygen nonstoichiometry at high temperature decrease in the order Ni > Co > Fe > Mn. However, the Co doped ceramic membranes exhibit higher oxygen permeation flux values than Ni-doped samples. For Ni- and Co-doped samples, bulk ionic transport controls the oxygen permeation rate.

INTRODUCTION

Mixed conductors exhibiting both oxide-ion and electronic conduction at high temperatures find several applications in electrochemical devices. For example, they can be used as cathodes in solid oxide fuel cells, oxygen separation membranes, and catalysts [1]. One of the most challenging applications of mixed conductors with high oxygen permeability is their utilization as oxygen separation membranes in reactors to partially oxidize light hydrocarbons like natural gas to value added products such as syngas and formaledyde [2]. In addition to the requirement of high electronic and ionic conductivity, this type of applications in reactors need materials that are structurally and chemically stable at high temperature and highly reducing environment.

High oxygen permeability for the perovskite oxides was first reported by Teraoka et al. [3] for the system $(La,Sr)(Co,Fe)O_{3-\delta}$. They also studied the effect of cation substitution on the oxygen permeation properties of this family of perovskite oxides [4]. Their results indicate that oxygen permeation flux increases with the A-site substitution in the order Sr < Ca < Ba and with B-site substitution in the order Mn < Cr < Fe < Co < Ni < Cu. Recently, we have reported the oxygen permeation properties and crystal structure stability of the n = 2 member of the Ruddlesden-Popper series with general formula $(La,Sr)_3(Fe,Co)_2O_{7-\delta}$ [5]. The crystal structure of these perovskite-related intergrowth oxides consists of double perovskite layers alternating with SrO rock salt layers along the c axis [6,7]. The substitution of Co for Fe in $Sr_{2.7}La_{0.3}Fe_{2-y}Co_yO_{7-\delta}$ increases both the electrical conductivity and the oxygen vacancy concentration at high temperature and thereby the oxygen permeability without changing the tetragonal symmetry of the crystal structure [5].

The objective of this paper is to investigate the effects of other cationic substitutions such as Ni and Mn for Fe on the oxygen permeation properties of $Sr_{2.7}La_{0.3}Fe_{2-y}M_yO_{7-\delta}$. We present here the synthesis, structural and chemical characterizations, electrical transport, and oxygen permeation measurements of the $Sr_{2.7}La_{0.3}Fe_{2-y}M_yO_{7-\delta}$ (M = Mn and Ni) system for $0 \leq y \leq 1.0$. The results are compared with those of other samples with M = Co.

EXPERIMENTAL

The $Sr_{2.7}La_{0.3}Fe_{2-y}M_yO_{7-\delta}$ (M = Mn, Co and Ni) oxides with $0.0 \leq y \leq 1.0$ were synthesized by a solution-based procedure [8]. Stoichiometric amounts of La_2O_3, $SrCO_3$, $Fe(CH_3COO)_2$, $Co(CH_3COO)_2.4H_2O$, $Ni(CH_3COO)_2.xH_2O$ and $Mn(CH_3COO)_2.4H_2O$ were dissolved in glacial acetic acid and refluxed for 2 h. Small amounts (2-10 mL) of H_2O and H_2O_2 were added, and the solution was refluxed further until it became clear. The solution was then heated in a hot plate to form a transparent gel. The gel was dried and decomposed at 400 °C for 10 min in air. The resulting powder was pressed into a disc and fired at 1400 °C for y = 0 and 0.6 and at 1350 °C for y = 1 in the case of M = Co and Ni and at 1500 °C for M = Mn. The densities of the sintered discs produced by the procedures were found to be > 90 % of the theoretical densities for M = Co and Ni.

X-ray powder diffraction data at room temperature were collected on a Phillips APD 3520 diffractometer using Cu Kα radiation. Rietveld refinements were carried out with the DBWS-9411 program [9]. Thermogravimetric analysis was performed with a Perkin-Elmer Series 7 Thermal Analyzer with a heating rate of 1°C/min in flowing air or N_2 atm. Total conductivity measurements in the temperature range 100°C ≤ T ≤ 940 °C were carried out in air using a four-probe dc technique with the Van der Pauw configuration.

Oxygen permeation measurements as a function of pO_2 at constant temperature and as a function of temperature at constant pO_2 gradients were performed in the temperature range 800 ≤ T ≤ 900 °C using a permeation setup that utilizes flowing He as a carrier gas and a Hewlett Packard 5880A gas chromatograph. The details of the oxygen permeation setup have been reported previously [8].

RESULTS AND DISCUSSION

An analysis of the X-ray diffraction patterns of $Sr_{2.7}La_{0.3}Fe_{2-y}M_yO_{7-\delta}$ samples with M = Mn and Ni in the range $0.0 \leq y \leq 1.0$ reveals that the Ni-containing samples are single phase while the Mn-containing samples contain a small amount of a perovskite phase. The crystal structures of all the samples were

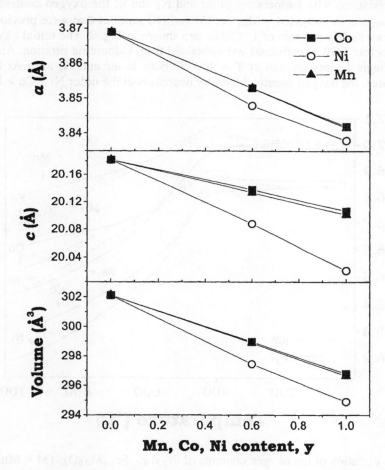

Figure 1: Variations of the lattice parameters and unit cell volume of air-annealed $Sr_{2.7}La_{0.3}Fe_{2-y}M_yO_{7-\delta}$ samples with Mn, Co, and Ni contents y.

refined at room temperature on the basis of the tetragonal space group I4/*mmm* using the Rietveld method. The solubility of M = Mn and Ni in $Sr_{2.7}La_{0.3}Fe_{2-y}M_yO_{7-\delta}$ is evident from the variations of the lattice parameters and unit cell volume with y (Fig. 1). Fig. 1 also includes the data of the Co-containing samples for comparison. The lattice parameters and unit cell volume decrease with increasing Co, Ni or Mn content due to the substitution of a smaller cation Mn^{4+}, $Co^{3+/4+}$ and Ni^{3+} for $Fe^{3+/4+}$ [10]. The substitution of Ni leads to larger changes compared to others.

The variations with temperature in air and N_2 atm of the oxygen contents of the $Sr_{2.7}La_{0.3}Fe_{1.4}M_{0.6}O_{7-\delta}$ (M = Mn, Fe, Co and Ni) samples that were previously cooled in air slowly at a rate of 1 °C/min are shown in Fig. 2. The initial oxygen contents before TGA experiments were obtained from iodometric titration. All the samples begin to lose oxygen at T ~ 400 °C both in air and N_2 atm. At high temperatures, the oxygen nonstoichiometry decreases in the order Ni > Co > Fe > Mn.

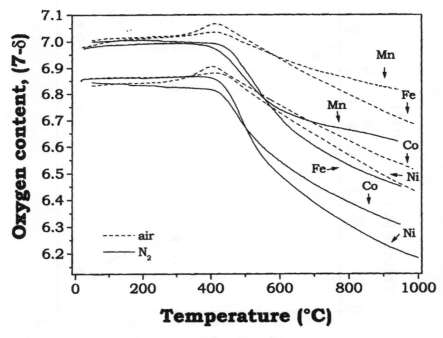

Figure 2: Variation of the oxygen contents of $Sr_{2.7}La_{0.3}Fe_{1.4}M_{0.6}O_{7-\delta}$ (M = Mn, Fe, Co, Ni) with temperature. Plots were obtained from TGA data collected in air (---) and N_2 (—) atm with a heating rate of 1 °C/min.

Materials for Electrochemical Energy Conversion and Storage

Fig. 3 shows the variations of the total conductivity σ of $Sr_{2.7}La_{0.3}Fe_{1.4}M_{0.6}O_{7-\delta}$ (M = Mn, Fe, Co and Ni) with temperature in air. All the samples show a thermally activated semiconducting behavior at low temperatures (T < 400 °C) with the conductivity increasing with T. After σ reaches a maximum at T ~ 400 °C, it begins to decrease due to a decreasing carrier concentration caused by oxygen loss [5]. For a given temperature, σ decreases in the order Ni > Co > Fe > Mn. The inset in Fig. 3 shows the variation of σ with Mn, Co or Ni contents at a constant temperature T = 900 °C in air. The σ values at high

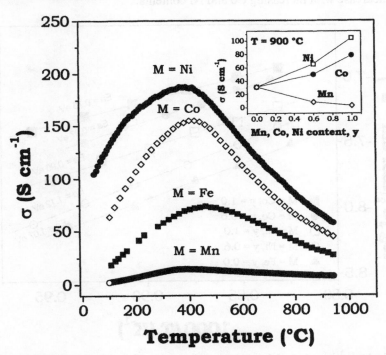

Figure 3: Variations of the total conductivity of $Sr_{2.7}La_{0.3}Fe_{1.4}M_{0.6}O_{7-\delta}$ (M = Mn, Fe, Co, Ni) with temperature. The inset displays the variations of the total conductivity at a constant T = 900 °C as a function of Mn, Co and Ni contents y.

temperature for the M = Fe, Co and Ni samples are above 30 S cm^{-1}. For Co-containing samples, it has been shown that the electrical conductivity is several orders of magnitude higher than the ionic conductivity [5], and we assume that it may also be the case for the Ni containing samples. Therefore, the oxygen permeation rates may be controlled by the oxide-ion conduction as in the case of the perovskite phases.

Fig. 4 shows the variations of the oxygen permeation flux jO_2 with inverse temperature for $Sr_{2.7}La_{0.3}Fe_{2-y}M_yO_{7-\delta}$ (M = Co, Ni) with $0 \leq y \leq 1.0$ at a constant pO_2 difference across the membrane. Measurements were performed with 1.5 mm thick samples with one side of the sample exposed to $pO_2' = 0.209$ atm (air) and the other side to a pO_2'' controlled by adjusting the He flow. While both the substitution of Co and Ni for Fe increases the jO_2 values, the Co-containing samples show higher jO_2 values compared to the Ni-containing samples. The apparent activation energy E_a values computed from the slope of the jO_2 vs $1/T$ curves decrease with increasing Co and Ni contents.

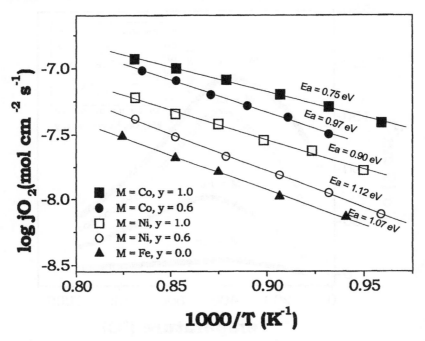

Figure 4: Arrhenius plot of the oxygen permeation flux (jO_2) for the $Sr_{2.7}La_{0.3}Fe_{2-y}M_yO_{7-\delta}$ samples at a constant pO_2 gradient $\log(pO_2'/pO_2'') = 2.2$. Measurements were conducted with 1.5 mm thick samples.

The lower jO_2 values of the Ni-containing samples compared to that of the Co-containing samples despite a higher conductivity and oxygen vacancy concentration at high temperature in the former could be due to slower oxygen surface exchange kinetics for the Ni-containing samples compared to the Co-containing samples or structural changes at high temperatures in the case of Ni-containing samples. No oxygen permeation measurements were performed on the Mn-containing samples due to the lower density obtained for these samples.

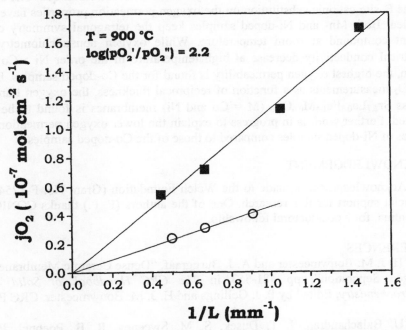

Figure 5: Variations of the oxygen flux with inverse sample thickness for $Sr_{2.7}La_{0.3}Fe_{1.4}M_{0.6}O_{7-\delta}$. ■: M = Co and ○: M = Ni.

To understand the process controlling the oxygen permeation properties of these compounds, we have prepared ceramic membranes with thicknesses $0.69 \leq L \leq 2.25$ mm for the Co-containing samples and $1.1 \leq L \leq 2.00$ mm for the Ni-containing samples. Fig. 5 shows the variations of jO_2 with reciprocal thickness for $Sr_{2.7}La_{0.3}Fe_{1.4}M_{0.6}O_{7-\delta}$ with M = Co and Ni. The experimental data was obtained at 900 °C at a constant pO_2 difference of $\log(pO_2'/pO_2") = 2.2$ across the membrane. The oxygen permeation flux shows a linear dependence on the reciprocal thickness with a good extrapolation to the origin for both the compositions. This kind of behavior is expected for materials in which the oxygen flux is bulk limited in accordance with the Wagner equation [5].

CONCLUSIONS

Perovskite-related intergrowth oxides $Sr_{2.7}La_{0.3}Fe_{2-y}M_yO_{7-\delta}$ (M = Mn and Ni) with $0 \leq y \leq 1.0$ have been synthesized by a solution-based technique. The effects of the B-site cationic substitution on the oxygen permeation properties have been studied. Both Mn- and Ni-doped samples keep the tetragonal symmetry of the parent compound at room temperature. While oxygen nonstoichiometry and electrical conductivity decrease at high temperature in the order Ni > Co > Fe > Mn, the highest oxygen permeability is found for the Co-doped samples. Based on jO_2 measurements as a function of reciprocal thickness, the oxygen transport across $Sr_{2.7}La_{0.3}Fe_{1.4}M_{0.6}O_{7-\delta}$ (M = Co and Ni) membranes is found to be bulk limited. Further work is in progress to explain the lower oxygen permeation flux values of Ni-doped samples compared to those of the Co-doped samples.

ACKNOWLEDGEMENT

Acknowledgment is made to the Welch Foundation (Grant No. F-1254) for financial support for this research. One of the authors (F. P.) thanks CONICET, Argentina, for a postdoctoral fellowship.

REFERENCES

[1]H. J. M. Bouwmeester and A. J. Burggraaf, "Dense Ceramic Membranes for Oxygen Separation"; pp 481-553 in *The CRC Handbook of Solid State Electrochemistry*, Edited by P. J. Gellings and H. J. M. Bouwmeester. CRC Press, 1997.

[2]U. Balachandran, J. T. Dusek, S. M. Sweeney, R. B. Poeppel, R. L. Mieville, P. S. Maiya, M. S. Kleefisch, S. Pei, T. P. Kobylinski, C. A. Udovich and A. C. Bose, "Methane to Syngas via Ceramics Membranes," *American Ceramic Society Bulletin*, **74**, 71-75 (1995).

[3]Y. Teraoka, H. M. Zhang, S. Furukawa, and N. Yamazoe, "Oxygen Permeation Through Perovskite-Type Oxides," *Chemistry Letters*, 1743-1746 (1985).

[4]Y. Teraoka, T. Nobunaga and N. Yamazoe, "Effect of Cation Substitution on the Oxygen Semipermeability of Perovskite Oxides," *Chemistry Letters*, 503-506 (1988).

[5]F. Prado, T. Armstrong, A. Caneiro, and A. Manthiram, "Structural Stability and Oxygen Permeation Properties of $Sr_{3-x}La_xFe_{2-y}Co_yO_{7-\delta}$ ($0.0 \leq x \leq 0.3$ and $0.0 \leq y \leq 1.0$)," *Journal of the Electrochemical Society* **148**, J7-J14 (2001).

[6]S. N. Ruddlesden and P. Popper, "The compound $Sr_3Ti_2O_7$ and its Structure," *Acta Crystallographica*, **11**, 54-55 (1958).

[7]S. E. Dann, M. T. Weller, and D. B. Curie, "Structure and Oxygen Stoichiometry in $Sr_3F_2O_{7-\delta}$," *Journal of Solid State Chemistry*, **97**, 179-185 (1992).

[8]T. Armstrong, F. Prado, Y. Xia, and A. Manthiram, "Role of Perovskite Phase on the Oxygen Permeation Properties of the $Sr_4Fe_{6-x}Co_xO_{13+\delta}$ System," *Journal of the Electrochemical Society*, **147**, 435-438 (2000).

[9]R.A. Young, A. Sakthivel, T. S. Moss, C. O. Paiva Santos, "DBWS-9411 program for Rietveld Refinement," *Journal of Applied Crystallography*, **28**, 366-367(1995).

[10]R. D. Shannon, "Revised Effective ionic radii and Systematic Studies of Interatomic Distances in Halides and Chalcogenides," *Acta Crystallographica* **A32**, 751-767 (1976).

Fuel Cells

LOW-COST MANUFACTURING PROCESSES FOR SOLID OXIDE FUEL CELLS

M.M. Seabaugh, B.E. McCormick, K. Hasinska, C.T. Holt,
S.L. Swartz and W.J. Dawson
NexTech Materials, Ltd.
720-I Lakeview Plaza Boulevard
Worthington, OH 43085, USA

ABSTRACT

NexTech Materials is developing low cost methods for producing a range of ceramic materials for solid oxide fuel cells (SOFCs). NexTech is developing low-cost fabrication methods for thin-film SOFCs that can be operated at low temperatures with inexpensive metallic interconnects. Dense electrolyte films are fabricated on porous electrode substrates from tailored aqueous nanoscale suspensions using aerosol spray coating methods. The nanoscale suspensions are prepared using low-cost hydrothermal synthesis methods. By controlling the dispersion chemistry and rheology, high-density green films can be deposited onto porous substrates. Multilayer planar components are being produced by tape casting of porous electrode (cathode or anode) substrates, electrolyte deposition and co-sintering, followed by deposition of opposite electrodes (anode or cathode) by screen printing. Both cathode and anode supported designs are being pursued, and fabrication methods for both configurations have been demonstrated.

INTRODUCTION

Fuel cells generate power by extracting the chemical energy of natural gas or other hydrogen-containing fuels without combustion. Advantages include high efficiency and very low release of polluting gases (e.g., NO_X) into the atmosphere. Of the various types of fuel cells, the solid oxide fuel cell (SOFC) can utilize hydrocarbon fuels without extensive external reforming, provides a source of high quality waste heat for cogeneration applications, and can be incorporated into power generation systems with extremely high efficiencies. However, the commercial viability of SOFCs, or any type of fuel cell, for widespread applications require that costs be reduced to levels competitive with existing power sources.

In this study, yttrium-stabilized zirconia (YSZ) has been chosen for the ceramic electrolyte. The fuel-side electrode (anode) material is a porous nickel-YSZ cermet, and the air-side electrode (cathode) material is a porous (La,Sr)MnO$_3$ (LSM) ceramic. It is generally accepted that the best way to reduce the cost of SOFC stacks is to reduce the operating temperature, which will allow the use of less expensive electrical interconnection materials (e.g., stainless steels). Lower operating temperatures also reduce balance-of-plant costs, especially those related to current collection and gas manifolding. One way to reduce operating temperature is to reduce the YSZ membrane thickness. By using supported thin-film electrolyte geometries, it should be possible to achieve high power densities at low temperatures and dramatically lower the manufacturing costs of SOFCs.

Current means of producing thin electrolyte layers can be expensive. Electrochemical vapor deposition (EVD) process can provide high-density and defect-free YSZ electrolyte films[1]. However, EVD is a capital intensive, batch process requiring specialized equipment and personnel. To lower the high costs of current SOFC fabrication, NexTech Materials has developed nanoscale YSZ electrolyte powders that can be deposited by aerosol spray methods and sintered to form dense electrolyte layers.

Traditional planar SOFC designs, based on relatively thick electrolyte membranes, operate at temperatures in the range of 950 to 1000°C to achieve optimum performance. Because of the high operating temperatures and the current use of external reformers, the materials, manufacturing and system costs are high ($1000/kW and higher), which limit the number of applications where SOFC power generation can be competitive. With lower manufacturing costs, SOFCs would become attractive options for several smaller-scale (<50 kW) power generation applications within various residential, transportation, industrial, and military market segments. SOFCs will become competitive for these applications if the stack cost can be reduced to the $100 per kilowatt level, which is about four times lower than is possible with current SOFC manufacturing technology.

The most effective approach for lowering this operating temperature is to reduce the electrolyte thickness. This is shown in Figure 1, where area-specific electrolyte resistance is plotted against temperature for a bulk YSZ electrolyte membrane, a bulk Sm-doped ceria (SDC) electrolyte membrane, and a thin-film YSZ membrane. If we consider only electrolyte resistance, replacing YSZ with SDC reduces the operating temperature to ~770°C, whereas reducing YSZ membrane thickness reduces the operating temperature to ~630°C. Of course, reduction of electrolyte thickness, by itself, is not sufficient to reduce operating temperature – the cathode and anode materials must also be engineered for low operating temperatures. By maintaining YSZ as the electrolyte material, however, only minor modifications to existing anode and cathode materials are needed to achieve lower temperatures, rather than a complete materials re-design that would be needed for alternative electrolyte materials.

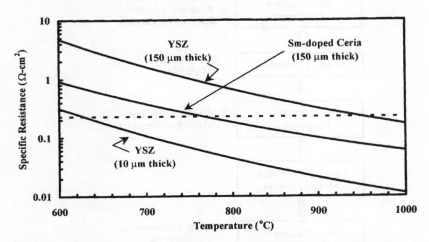

Figure 1. Effect of temperature on specific resistance of bulk YSZ and Sm-doped ceria (SDC) electrolyte membranes, and thin-film YSZ electrolyte membranes.

EXPERIMENTAL

To lower the cost of electrolyte deposition, a colloidal method of depositing nanoscale YSZ on porous substrates has been designed. The core of this deposition technology is the hydrothermally produced YSZ powder, which is made by the process shown in Figure 2. The product powder after hydrothermal reaction has a crystallite size of 5-15 nm (Figure 3). By controlling the surface chemistry of the powder, concentrated suspensions (>40 wt% solids) can be obtained with particle size distributions with a median particle size of 100 nm (Figure 4). The suspensions have a relatively low viscosity and easily penetrate the pores of substrates during deposition; organic viscosity and surface-chemistry modifiers have been used to tailor the wetting and cohesion of the film to allow continuous films to be deposited with a minimum of wicking.

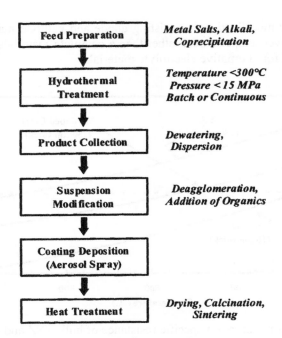

Feed Preparation	*Metal Salts, Alkali, Coprecipitation*
Hydrothermal Treatment	*Temperature <300°C Pressure <15 MPa Batch or Continuous*
Product Collection	*Dewatering, Dispersion*
Suspension Modification	*Deagglomeration, Addition of Organics*
Coating Deposition (Aerosol Spray)	
Heat Treatment	*Drying, Calcination, Sintering*

Figure 2. Nanoscale spray suspension preparation and application process.

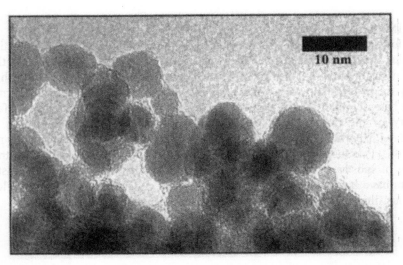

10 nm

Figure 3. TEM micrograph of hydrothermally-derived YSZ crystallites.

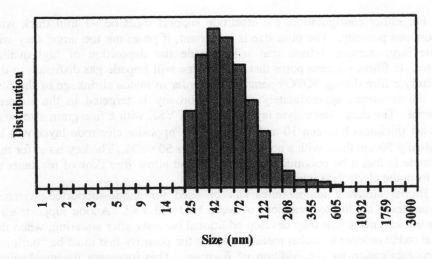

Figure 4. Particle size distribution of hydrothermally derived YSZ suspension.

The suspension described above provides an excellent raw material for applying electrolyte films. The nanoscale oxides used in the process can be tailored to densify at targeted temperatures between 1200 and 1400°C by manipulating the surface area of the spray solution. As the surface area of the powder increases, the energy available for densification increases, and lower sintering temperatures are viable. However, high surface area often leads to increased shrinkage of the films and cracking during sintering. To take advantage of the lower processing temperatures offered by the nanoscale suspensions, the film sintering shrinkage must be accommodated by the substrate. Applying the film to an unsintered substrate offers such an opportunity. By matching film and substrate shrinkage, co-sintering below 1350°C is plausible.

In order to achieve the performance and cost goals, fabrication methods must be developed for thin-film SOFCs to take advantage of low operating temperatures that allow inexpensive metallic interconnects. For the current study, the cell design has three oxide layers: support electrode, electrolyte film, and opposite electrode. The support electrode layer can be either the cathode or anode; each configuration has advantages and disadvantages. Anode-supported thin-film SOFCs with high power densities and low operating temperatures have been demonstrated [2-4]. A particular advantage of the anode-supported approach is that the NiO/YSZ anode and YSZ electrolyte materials do not react during co-sintering. However, the cathode-supported approach offers lower raw materials costs and the potential for longer-term reliability (due to better thermal expansion match between support electrode and deposited electrolyte film).

For either configuration, the electrode support must be ~1 mm thick with continuous porosity. The pore size is important; if pores are too large, they will create large surface defects that will impede the deposition of high-quality electrolyte films, whereas pores that are too fine will impede gas diffusion to the electrolyte film during SOFC operation. In order to match shrinkage of the film and the substrate, approximately 40 vol% porosity is targeted in the sintered material. The dense electrolyte layer consists of YSZ with a fine grain size, with a target thickness between 10 and 20 μm. The opposite electrode layer will be nominally 50 μm thick with a porosity value near 50 vol%. The key issue for this electrode is that it be continuous, conductive and allow free flow of reactants to and from the electrolyte interface.

Higher sintering temperatures can be used for anode-supported systems, because there are limited reactions between NiO and YSZ. Anode supports also have the advantage that they develop additional porosity after sintering, when the nickel oxide reduces to nickel metal, lowering the porosity that must be "built in" during fabrication by the addition of fugitive. This increases the mechanical strength of the substrates after fugitive removal. For this study, anode supports were fabricated at Oak Ridge National Laboratory by tape casting and lamination. The substrates were supplied in 2.5 cm square coupons, ~1 mm thick. The green samples were calcined to remove the binder and open the pore structure. A nanoscale spray solution was prepared by the method shown in Figure 2, and deposited in a single coating step. An aerosol spray nozzle was used to deposit the suspension onto the surface of planar substrates. The coated substrate was then sintered at 1350°C for one hour, to produce a bilayer with the microstructure shown in Figure 5. As can be seen, highly dense and continuous YSZ film layers can be achieved with 20 μm thickness.

While the chemical compatibility of the anode and electrolyte allows high temperature sintering, the nickel reduction step may increase the likelihood of mechanical failures on large surface-area elements. The reliability of anode-supports under temperature cycling is complicated by the thermal expansion mismatch between the nickel-YSZ anode and the YSZ electrolyte film. Because of these concerns, a major advantage of cathode supported design is that the LSM cathode and YSZ electrolyte material have an almost perfect expansion match, making improved mechanical reliability a key advantage of a cathode-supported approach. For the cathode-supported approach, the most significant obstacle to success is matching the sintering shrinkage curve of the LSM and YSZ. This task is further complicated by the need to minimize the processing temperature, in order to avoid adverse reactions (lanthanum zirconate formation and/or manganese diffusion) at the cathode/electrolyte interface.

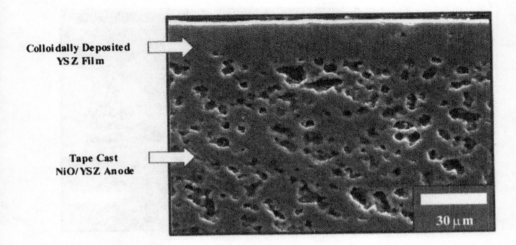

Colloidally Deposited YSZ Film

Tape Cast NiO/YSZ Anode

30 μm

Figure 5.   Anode supported YSZ electrolyte film. Anode
substrate prepared at Oak Ridge National Laboratory.

The cathode-supported approach also was demonstrated.   Porous LSM cathode substrates were made using tape casting and lamination methods. LSM powder was mixed with an organic fugitive material in a tape casting slurry, and then cast at a thickness of 200 μm. The dried tape was punched, laminated and calcined. During the calcination process, both the organic binder phase and fugitive decomposed, and the fine LSM powder densified sufficiently to allow easy handling of the substrates. To prevent chemical interaction at high sintering temperatures, an interfacial layer was applied to the cathode supports, using a colloidal deposition route similar to that used for the electrolyte. The interfacial layer also increases the effective electrode/electrolyte interface area by creating a larger number of fine electrode/electrolyte interfacial contacts.   Both the interlayer and electrolyte are applied using aerosol colloidal deposition, and then co-sintered with the support. The tri-layer sample was sintered at 1350°C for one hour. A micrograph of a cathode supported YSZ film with interlayer is shown in Figure 6. The interlayer provides a distinct layer separating the LSM layer (with large interconnected pores) from the electrolyte. As shown, electrolyte thickness can be as thick as 30 μm. For the targeted application, the membrane thickness will be reduced to 10-20 μm.

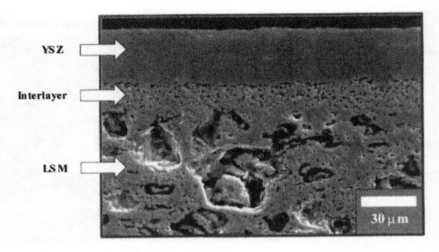

Figure 6. Cathode supported YSZ electrolyte film with interfacial layer.

SUMMARY AND FUTURE WORK

Colloidal deposition and sintering of nanoscale YSZ suspensions have been demonstrated for fabricating 10-20 mm thick electrolyte membranes for multi-layer, supported thin-film fuel cell configurations. Both anode and cathode supports have been successfully fabricated and coated at temperatures at or below 1400°C. Future work is directed toward reducing electrolyte film thickness, reducing sintering temperature, and beginning testing of cells produced using these methods.

ACKNOWLEDGMENTS

NexTech Materials gratefully acknowledges the financial support of the U.S. Department of Energy under Contract No. DE-AC26-00NT40706 and the State of Ohio under Contract No. TECH-00-089. The authors also wish to thank Dr. Tim Armstrong of Oak Ridge National Laboratory for supplying anode substrates used to make the sample shown in Figure 5.

REFERENCES

[1]U.B. Pal and S.C. Singhal, J. Electrochem.Soc., 137, 2937 (1990)
[2]S. de Souza, et al., Solid State Ionics **98**, 57-61 (1997).
[3]J.W. Kim, et al., Journal of the Electrochemical Society **146**, 69-78 (1999).
[4]H. Ohrui, et al, Journal of Power Sources **71**, 185-89 (1998).

MANUFACTURING ROUTES AND STATE-OF-THE-ART OF THE PLANAR JÜLICH ANODE-SUPPORTED CONCEPT FOR SOLID OXIDE FUEL CELLS

W. A. Meulenberg, N. H. Menzler, H. P. Buchkremer and D. Stöver

Forschungszentrum Jülich GmbH,
Institute for Materials and Processes in Energy Systems, IWV-1
52425 Jülich, Germany

ABSTRACT

Solid oxide fuel cells (SOFCs) are electrochemical devices for the direct conversion of chemical fuel energy into electricity. At Forschungszentrum Jülich, an advanced planar SOFC has been developed. It is characterised by a thick anode substrate and a thin electrolyte. The supporting 250 mm × 250 mm anode substrate, a 1.5 mm thick porous cermet of nickel and yttria-stabilised zirconia is made by a special warm pressing method. A thin fine-structured anode functional layer is deposited by vacuum slip casting. The same deposition method is used for the yttria-stabilised electrolyte with a sintered thickness of only 5-10 μm. The double-layered porous cathode with a total thickness of 50 μm is made from $La_{0.65}Sr_{0.3}MnO_3$ (LSM). This perovskite-type material is deposited using the wet powder spraying® process. Other alternative processing methods like tape casting for the substrate and screen printing for the anode functional layer, electrolyte and cathode are under investigation to select a cost-effective manufacturing route for all components.

INTRODUCTION

In the fuel cell, a direct conversion of chemical into electrical energy takes place without any combustion process. This enables a more efficient and environmentally acceptable use of our energy resources. Energy is produced in a fuel cell stack by electrochemical reaction between a fuel gas (e.g. hydrogen) and atmospheric oxygen. During this reaction, oxygen ions from the air are transported from the cathode side through an ion-conducting electrolyte to the anode side, where the oxygen reacts with the hydrogen of the fuel gas to form water. During the recombination of hydrogen and oxygen into water, electrons are released which can be used for the electricity supply of a final consumer.

The fabrication routes for the individual cell components of the different SOFC designs (planar - tubular) differ greatly depending on which cell layer is to perform the supporting function in the cell. Whereas for the planar concept in most cases the anode or the electrolyte ensures the mechanical stability of the cells, in the tubular concepts the cathode is often the supporting component[1,2].

In the anode substrate concept the anode is the supporting component of the cell and must therefore display sufficient mechanical stability. The substrates are predominantly produced by tape casting[3,4,5,6,7,8,9] or using a pressing process[4,10,11]. As a rule, an anode functional layer a few micrometers in thickness is then deposited on the substrate by vacuum slip casting[11] or screen printing[9]. In the planar concepts where the anode has no supporting function it is frequently deposited by screen printing or plasma spraying[12].

In the tubular fuel cell system in most cases the cathode is the supporting cell component. The cathode tubes are produced using an extrusion process[13,14]. For the other concepts, the cathodes are produced by wet powder spraying[15], screen printing[3].

In the planar system the electrolyte is deposited mostly by slip casting[15,16], vacuum slip casting[17,18], screen printing[19,20], plasma spraying[21] and wet powder spraying. State-of-the-art in the tubular system is electrochemical vapour deposition[14,22] and plasma spraying. Nowadays, research focuses on the replacement of cost-intensive techniques, i.e. plasma spraying and electrochemical vapour deposition, by simple and cost-effective procedures like slip casting, vacuum slip casting and screen printing.

At Forschungszentrum Jülich, the so-called anode-supported concept has been developed in which an electrolyte of ZrO_2 and a cathode of lanthanum strontium manganite is deposited onto an anode of Ni/ZrO_2 about 1.5 mm in thickness.

Figure 1:   Solid oxide fuel cells at different manufacturing stages.

Figure 1 shows the cells in different manufacturing states . At the top (from left to right) an anode substrate and an anode substrate with deposited electrolyte are shown. At the bottom of Figure 1 the completed cells with the black cathode layer are shown.

MANUFACTURING ROUTES OF THE JÜLICH CELLS

For the manufacturing of the porous anode substrate the two routes of warm pressing and tape casting are used.

The manufacturing of the anode substrate by warm pressing (Figure 1) is effected with a special coat-mix® powder, where the surfaces of the NiO and the 8 mole % yttria-stabilised zirconia (8YSZ) are coated with a resin.

At first the resin will be dissolved in ethanol at a temperature of 50-60°C. After this the powder mixture of NiO and 8YSZ is given into the binder suspension. The formed suspension will be cooled down and given into acidic water. The water will then be mixed with the suspension, and the binder precipitates. The powder surface is coated with the binder. After this the powder is filled into a pressing die. The pressing process was carried out at 120°C and 1 MPa.

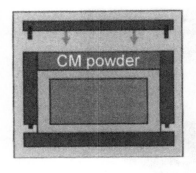

Figure 2: Substrate manufacturing by the warm pressing process with coat-mix® powder.

At this temperature, the binder is initially soft and combines the powder particles with each other. It then begins to cure by splitting water off the resin. The pressing die can be heated or cooled by oil to shorten the pressing cycle. The combustion of the binder and the sintering process take place in one step at a sintering temperature between 1200°C and 1300°C producing the open porosity required for the anode substrates. After debindering of the finished fuel cell, the pressed substrate has a thickness of 1.5 mm.

The alternative route for substrate manufacturing at Forschungszentrum Jülich is the tape casting process which is shown in Figure 3.

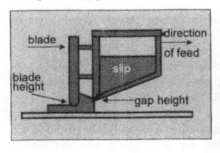

Figure 3: Anode processing by the tape casting process.

The tape casting process is a very cost-effective process, because a large quantity of substrates can be produced in a short time by casting a slip onto a plastic tape as shown in Figure 3. For slip production, an alcoholic solvent is first mixed with the NiO and 8YSZ powder and a dispersant and the mixture is ground. Subsequently, the binder and the plastifier are added and mixing is repeated. In order to establish the porosity required for the the anode substrate, as a rule, graphite or a resin is added. After drying, a plastic tape is obtained which is easy to handle. Graphite or resin and the binder leave pore structures during burning out through which the fuel gas can flow. In the final state, the tape has a thickness between 300 and 500 μm.

The next step in the anode-supported SOFC concept is the coating of the substrate with an anode functional layer and an electrolyte. The functional layer (AFL) is composed of 50 wt.% NiO and 50 wt.% 8YSZ. The electrolyte (El) is based on fine 8YSZ powder. Both layers were deposited by vacuum slip casting (Figure 4).

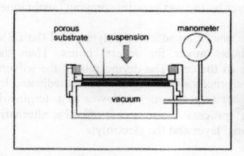 

Figure 4: Anode functional layer and electrolyte manufacturing by vacuum slip casting.

In this technique the layer suspension (ethanol, zirconia and polyethylene imine) is deposited on the top side of the substrate while on the opposite side vacuum is drawn. The thickness of the layer (AFL: 5 μm, El: 5 μm) is adjusted by the amount of suspension. During coating the ethanol migrates through the porous substrate and a layer is formed on the material. After coating the substrate is dried and sintered.

The double-layered porous cathode with a total thickness of 50 μm is made from $La_{0.65}Sr_{0.3}MnO_3$. This spray-dried and calcinated perovskite-type material is deposited using the wet powder spraying® process which is shown in Figure 5.

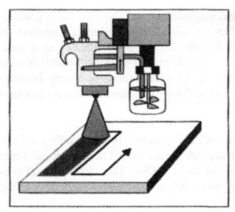

Figure 5: Processing of the cathode layer by the wet powder spraying® technique.

The suspension of the carrier liquid (ethanol), an adhesive polymer and the LSM powder were homogenized in a tubular mixer for several hours. Then the suspension is sprayed onto the surface of the cell. The drying rate of the solvent has a substantial influence on the homogeneity of the layer. It can be adjusted by adding an additional solvent with higher (e.g acetone) or lower (e.g. terpineol) drying rate. The wet powder spraying® process is also a cost-effective alternativ route to manufacture the anode functional layer and the electrolyte.

An additional route for the manufacturing of thin electrodes and the electrolyte is the screen printing method (Figure 6).

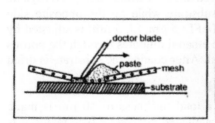

Figure 6: Layer manufacturing by screen printing.

In screen printing, a paste is prepared from a solvent, binder and starting powder. A homogeneous layer can be deposited by selecting a suitable mesh width of the screens. In addition to the grain size of the starting particles, the mesh width in printing has also an influence on the layer structure.

Figure 7 shows the microstructure of a reduced solid oxide fuel cell.

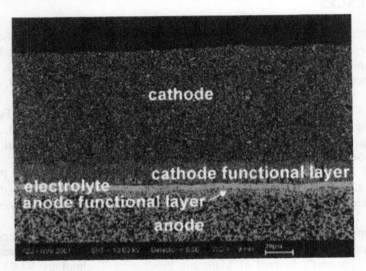

Figure 7: Cross-section of a reduced single cell.

The picture shows the layer structure of a solid oxide fuel cell produced by the standard route. The dense electrolyte layer 5 μm in thickness is placed between the two porous electrodes consisting of two layers each with the thinner layers next to the electrolyte being finer-grained. Studies concerning the performance of the cathode have shown that the microstructure of the cathode is of decisive significance[23,24]. It can be selectively adjusted by varying the manufacturing parameters in the different processes and the sintering temperature.

SUMMARY

The manufacturing routes used at Forschungszentrum Jülich for the single cell components of the planar anode-supported solid oxide fuel cell concept are shown. The standard substrates are made by a warm pressing process but also tape casting is a cost-effective alternative. The substrates are deposited with an anode functional layer and an electrolyte using vacuum slip casting. The cathode is produced by the wet powder spraying® technique. Wet powder spraying and screen printing are possible low-cost processes for the fabication of the thin anode functional layer, the electrolyte and the cathode layers.

REFERENCES

[1] H. P. Buchkremer, U. Diekmann, L. G. J. de Haart, H. Kabs, U. Stimming, D. Stöver, "Advances in the anode supported planar SOFC technology", Proc. 5th Int. Symp. Solid Oxide Fuel Cells (SOFC-V), eds.: U. Stimming, S. C. Singhal, H. Tagawa, W. Lehnert, The Electrochemical Society, Pennington, NJ, 1997, p. 160

[2] S. C. Singhal, "Recent Progress in Tubular Solid Oxide Fuel Cell Technology", Proc. 5th Int. Symp. Solid Oxide Fuel Cells (SOFC-V), eds.: U. Stimming, S. C. Singhal, H. Tagawa, W. Lehnert, The Electrochemical Society, Pennington, NJ, 1997, p. 37

[3] G. M. Christie, J. P. Ouweltjes, R. C. Huiberts, E. J. Siewers, F. P. F. van Berkel, J. P. P. Huijsmans, "Development and Manufacturing of SOFC Components at ECN, the Netherlands", 3rd International Fuel Cell Conference, 30.11-03.12.1999 Nagoya, Japan, (1999), 361-364

[4] D. Simwonis, H. Thülen, F. J. Dias, A. Naoumidis, D. Stöver, "Properties of Ni/YSZ porous cermets for SOFC anode substrates prepared by tape casting and coat-mix® process", Journal of Materials Processing Technology 92-93 (1999), 107-111

[5] C. Lutz, A. Roosen, D. Simwonis, A. Naoumidis, H. P. Buchkremer, Foliengießen eines porösen Anodensubstrats für die Hochtemperaturbrennstoffzelle" Volume III, Werkstoffwoche Munich 1998, Walter Muster, Rainer Link, Josef Ziebs (editors), 1st edition, Wiley VCH; Weinheim, New-York, Chichester, Brisbane, Singapore, Toronto, (1999), 155-160

[6] S. Primdahl, M. J. Jørgensen, C. Bagger, B. Kindl, "Thin Anode Supported SOFC", Solid Oxide Fuel Cells VI, 17-22 Oct. 1999 in Honolulu, Hawaii, USA, Electrochemical Proceedings Volume 99-19, (1999) 793-799

[7] D. Gosh, G. Wang, R. Brule, E. Tang, P. Huang "Performance of Anode Supported Planar SOFC Cells" Solid Oxide Fuel Cells VI, 17-22 Oct. 1999 in Honolulu, Hawaii, USA, Electrochemical Proceedings Volume 99-19, (1999) 822-829

[8] J. P. Ouweltjes, F. P. F. van Berkel, P. Nammensma, G. M. Christie "Development of 2nd Generation, Supported Electrolyte, Flat SOFC Components at ECN", Solid Oxide Fuel Cells VI, 17-22 Oct. 1999 in Honolulu, Hawaii, USA, Electrochemical Proceedings Volume 99-19, (1999) 803-811

[9]Q. N. Minh, T. R. Amstrong, J. R. Esopa, J. V. Guiheen, C. R. Horne, J. Van Ackeren, "Tape-calendared monolithic and flat plate Solid Oxide Fuel Cells", Proceedings of the 3nd International Symposium on Solid Oxide Fuel Cells, 16-21 May 1993 in Honolulu, Hawaii, Volume 93-4, (1993), 801-808

[10]D. Simwonis, F. J. Dias, A. Naoumidis, D. Stöver, "Manufacturing of Porous Cermet Substrates for Solid Oxide Fuel Cells by Coat-Mix Process", Proc. 5th Eur. Conf. Advanced Materials, Processes and Applications (EUROMAT '97), eds.: L. A. J. L. Sarton, H. B. Zeedijk, Netherlands Society for Materials Science, Zwijndrecht, Netherlands, (1997), Vol. 2, p. 375

[11]D. Stöver, U. Diekmann, U. Flesch, H. Kabs, W. J. Quadakkers, F. Tietz, I. C. Vinke, "Recent Developments in Anode Supported Thin Film SOFC at Research Centre Jülich", Solid Oxide Fuel Cells VI, 17-22 Oct. 1999 in Honolulu, Hawaii, USA, Electrochemical Proceedings Volume 99-19, (1999) 813-821

[12]G. Schiller, R. Henne, M. Lang, S. Schaper, "Development of Metallic Substrate Supported Thin-Film SOFC by Applying Plasma Spray Techniques" Solid Oxide Fuel Cells VI, 17-22 Oct. 1999 in Honolulu, Hawaii, USA, Electrochemical Proceedings Volume 99-19, (1999) 893-903

[13]S. C. Singhal, "Status of Solid Oxide Fuel Cell Technology", Proceedings of the 17th Risø International Symposium on Materials Science, Roskilde, Denmark, (1996), 123-138

[14]S. E. Veyo, "Tubular SOFC field unit status", 3rd International Fuel Cell Conference, 30.11-03.12.1999 Nagoya, Japan, (1999), 327-332

[15]T. Nakayama., H. Tajiri, K. Hiwatashi, H. Nishiyama, S. Kojima, M. Aizawa, K. Eguchi, H. Arai „Performance characteristics of a tubular SOFC by wet process" Proc. of the 5th Int. Symposium on Solid Oxide Fuel Cells (SOFC-V) (1997), 187-195

[16]C. Wang, W. L. Worrell, S. Park, J. M. Vohs, R. J. Gorte „Fabrication and performance of thin-film YSZ solid oxide fuel cells between 600 and 800°C" Proc. of the 6th Int. Symp. on Solid Oxide Fuel Cells (SOFV-VI), Singhal S.C., Dokiya M. (Eds.), Oct. 17th-22nd 1999, Honolulu, USA, (1999), 851-860

[17]P. Batfalsky, H. P. Buchkremer, D. Froning, F. Meschke, H. Nabielek, R. W. Steinbrech, F. Tietz „Operation and analysis of planar SOFC stacks" Proc. of the 3rd Int. Fuel Cell Conf. Nov 30th – Dec 3rd 1999, Nagoya, Japan, (1999), 349-352

[18]D. Ghosh, G. Wang, R. Brule, E. Tang, P. Huang „Performance of anode supported planar SOFC cells" Proc. of the 6th Int. Symp. on Solid Oxide Fuel Cells (SOFV-VI), Singhal S.C., Dokiya M. (Eds.), Oct. 17th-22nd 1999, Honolulu, USA, (1999), 822-829

[19]G.-Y. Kim, S.-W. Eom, S.-I. Moon „Cell properties of SOFC prepared by doctor blade and screen printing method" Proc. of the 5th Int. Symposium on Solid Oxide Fuel Cells (SOFC-V), (1997), 700-709

[20] G. M. Christie, P. Nammensma, J. P. P. Huijsmans, „Status of anode supported thin electrolyte ceramic SOFC compositions developed at ECN" Proc. of the 4th Europ. Solid Oxide Fuel Cell Forum, McEvoy A. J. (Ed.), July 10th-14th 2000, Lucerne, Switzerland, (2000), 3-11

[21] G. Schiller, R. Henne, M. Lang, R. Ruckdäschel, S. Schaper „Fabrication of thin-film SOFC by plasma spray techniques" Proc. of the 4th Europ. Solid Oxide Fuel Cell Forum, A. J. McEvoy (Ed.), July 10th-14th 2000, Lucerne, Switzerland, (2000), 37-46

[22] S. C. Singhal „Progress in tubular solid oxide fuel cell technology" Proc. of the 6th Int. Symp. on Solid Oxide Fuel Cells (SOFV-VI), Singhal S.C., Dokiya M. (Eds.), Oct. 17th-22nd 1999, Honolulu, USA, (1999), 39-51

[23] A. Ahmad-Khanlou, „Alternative Werkstoffe für Komponenten der Hochtemperaturbrennstoffzelle (SOFC) zur Herabsetzung der Betriebstemperatur", PhD thesis at Research Centre Jülich, Jül 3797 (2000)

[24] K. Sasaki, J. P. Wurth, R. Gschwend, M. Gödickemeier, L. J. Gauckler, Journal Electrochem. Soc., 143, (1996), pp. 530

MATERIALS AND MICROSTRUCTURES FOR IMPROVED SOLID OXIDE FUEL CELLS

S. Huss, R. Doshi, J. Guan, G. Lear, K. Montgomery, N. Minh, and E. Ong
Honeywell Engines and Systems
2525 W. 190th Street
Torrance, CA 90504-6099

ABSTRACT

Continuing efforts to improve the performance of SOFCs have included extensive fabrication process development and engineering of cell components (anode, electrolyte and cathode). Improvements in electrolytes result from the development of processing methods to fabricate thin-film electrolytes with reduced resistance. Significant improvements in cell performance are achieved by increasing the porosity and optimizing the microstructure of the electrodes. This paper describes efforts to develop processing methods for thin-film electrolyte fabrication and improved electrode microstructures to reduce polarization losses and enhance cell performance.

INTRODUCTION

While significant advances in SOFC technology have been made in the past several years, several challenges have yet to be resolved before the SOFC becomes a viable and marketable energy conversion product. These challenges include improved performance, longer life and lower cost.

Figure 1 illustrates the voltage drop of the three principal components of the fuel cell (anode, electrolyte, and cathode); all three contribute to performance losses, particularly at these temperatures. Since all three components contribute to overall loss, improvements to each are valuable for overall cell performance. In the following sections, techniques and strategies for augmenting the electrolyte, cathode and anode performance in a planar fuel cell configuration at reduced temperatures (<800°C) are described. The exact configuration of the fuel cell design discussed herein is described elsewhere [1]. The cell itself is an anode supported, thin electrolyte cell. The anode and electrolyte are fabricated using a tape calendering technique described below. The cathode is applied using a screen printing technique.

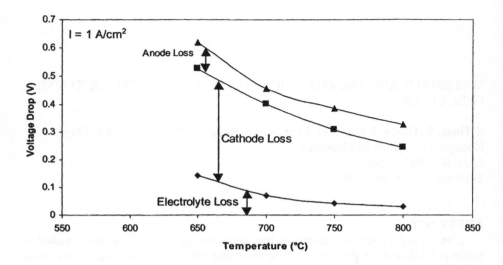

Figure 1. Performance loss contributions of the key fuel cell components as a function of temperature.

REDUCING ELECTROLYTE LOSSES AT REDUCED TEMPERATURES

Much effort has been made to reduce SOFC operating temperatures from the typical operating range of 900-1000°C to 800°C and below. Reduced operating temperatures would facilitate the use of metal interconnect materials. The potential benefits of metal interconnects include improved manufacturability and lower cost. Since the electrolyte conductivity (and ultimately the fuel cell performance) falls dramatically with temperature, higher conductivity electrolyte materials and/or thinner electrolytes must be developed to compensate for the performance losses associated with the temperature drop.

One obvious method to reduce ohmic losses for a given electrolyte composition is to reduce its thickness. Figure 2 illustrates the effects of electrolyte thickness on the area-specific resistance (ASR) of the cell. The data also show significant increases in the cell ASR with decreasing temperature. This increase is due to both a reduction in ionic conductivity of the electrolyte and an increase in the polarization of both electrodes. Thus, reducing the electrolyte thickness is a simple approach to minimize the ohmic losses for efficient operation at relatively low temperatures.

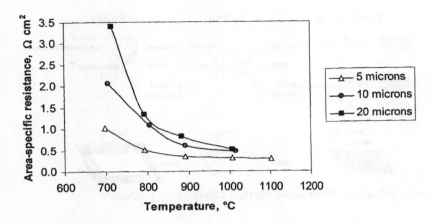

Figure 2. Area-specific resistance data as a function of temperature for three thin-film electrolyte fuel cells of varying electrolyte thickness.

Honeywell has developed a novel low-cost processing technique based on tape calendering to fabricate thin electrolyte cells [2]. The technique, known as tape-calendering is shown schematically in Figure 3. In this process, electrolyte and support electrode (e.g. anode) tapes are first made by combining the constituent ceramic powders and organic binders in a high shear mixer. These tapes are then rolled together into a bilayer using a two-roll mill. Rolling successive layers of electrode tape with the existing bilayer progressively thins the electrolyte. The final bilayer is fired at elevated temperatures to remove organics and sinter the ceramics. The cathode layer is then applied to the electrolyte by such a method as screen printing. Figure 4 shows a micrograph of a typical cell made by tape calendering. As seen in the figure, the yttria-stabilized zirconia (YSZ) electrolyte is continuous and fully dense (approximately 3-4 microns thick in this case). Tape-calendered cells have shown excellent performance at 800°C and below. Figure 5 shows the peak power density of more than 1 W/cm^2 at 800°C and nearly 450 mW/cm^2 at 650°C for a thin YSZ electrolyte cell fabricated by the tape-calendering process.

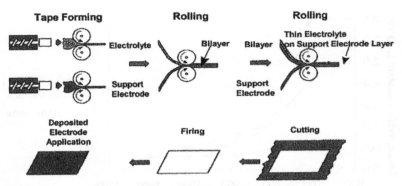

Figure 3. Schematic diagram of the tape calendering process.

Figure 4. SEM micrograph of the crossection of a fuel cell fracture surface showing the dense electrolyte between the porous anode and cathode layers.

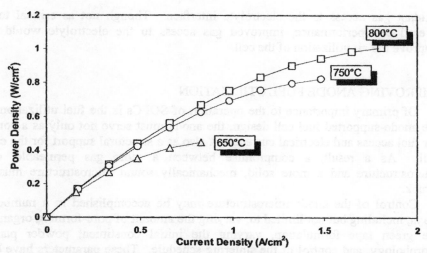

Figure 5. Power density as a function of temperautre for a typical tape-calendered SOFC.

IMPROVING CATHODE PERFORMANCE

Figure 1 above illustrates that the cathode is the dominant contribution to cell polarization losses. Consequently, efforts to improve overall cell performance should consider strategies to improve cathode performance.

The cathode must provide sufficient oxidant access to the electrolyte surface while maintaining good interfacial contact with the electrolyte and good conductive pathways for electrical current. Near the electrolyte interface, the electrochemical properties of the cathode material and the distribution of catalytic sites are of primary importance. Further away from the interface, adequate oxidant access through the cathode is important. Currently, a porous composite of ionically conductive material, typically yttria stabilized zirconia (YSZ), and an electronically conductive material, typically lanthanum strontium manganate (LSM), is used to enhance cathode activities. The performance of this composite has been shown to exceed that of cathodes made from LSM alone [3]. As an alternative to a composite, the use of a single mixed conductor such as a cobaltite material might improve the overall efficiency of the process. On the other hand, the spatially varying requirements of the cathode layer suggest that graded structures may improve the cathode's performance. A layered cathode structure with varying catalytic activity as a function of distance from the electrolyte surface might enhance performance.

In addition to exploring these alternative cathode compositions, Honeywell continues to explore microstructural engineering as a means to provide superior oxidant gas access to the electrolyte interface. Though not as critical to the overall cell performance, improved gas access to the electrolyte would also improve the air utilization of the cell.

IMPROVING ANODE FUEL UTILIZATION

Of primary importance to the operation of SOFCs is the fuel utilization. In the anode-supported fuel cell design, the anode must serve not only as a conduit for fuel access and electrical current, but also as a structural support for the entire cell. As a result, a compromise between a more gas permeable, open microstructure and a more solid, mechanically sound microstructure must be struck.

Control of the anode microstructure may be accomplished in a number of ways, including but not limited to, varying the amount of pore-forming organic in the green tape formulation, varying the initial constituent powder particle morphology, and control of the sintering schedule. These parameters have been modified to create a proprietary, engineered anode structure with more open fuel access to the electrolyte. The difference between the baseline and engineered anode structures is shown schematically in Figure 6.

The more open anode structure has demonstrated significantly improved performance over the baseline. The difference is more readily seen with dilute fuel gases under conditions of high fuel utilization. Figure 7 shows the polarization curves for a fuel cell with the engineered anode and a fuel cell with the baseline anode. In this figure, power density curves are given for performance in a synthetic fuel gas, or syngas, containing 19% H_2, 24% CO, 1% CO_2, and a balance N_2. The power density achieved with the engineered anode cell is nearly twice that of its baseline counterpart at a 20% higher fuel utilization.

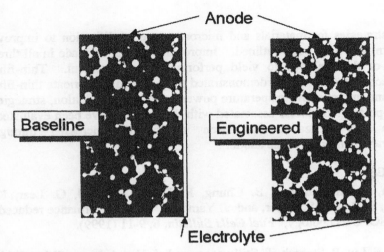

Figure 6. Schematic illustration of the cross-section of baseline and engineered anode structures.

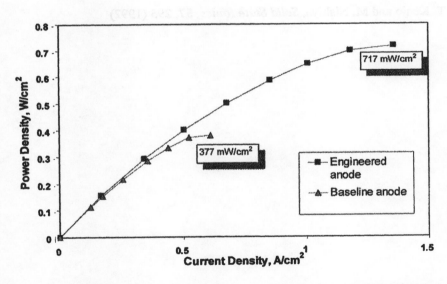

Figure 7. Polarization curves for cells with the engineered and baseline anodes. The fuel utilizations in syngas (containing 19% H_2, 24% CO, 1%CO_2, balance N_2) are 60% and 40% for the engineered anode and baseline anode, respectively.

SUMMARY

Several strategies for materials and microstructure modification to improve SOFC performance have been outlined. Improvements can be made in all three components of the fuel cell to yield performance improvement. Thin-film processing techniques have been demonstrated to successfully fabricate thin-film electrolytes and improve low temperature power densities. In addition, strategies for reducing polarization losses associated with the cathode have been evaluated. Finally, the fuel utilization has been significantly increased through microstructural engineering of the anode.

REFERENCES

[1] N. Minh, A. Anumakonda, B. Chung, R. Doshi, J. Ferral, G. Lear, K. Montgomery, E. Ong, L. Schipper, and J. Yamanis, "High-performance reduced-temperature SOFC technology," *Fuel Cells Bulletin*, 6, 9-11 (1999).

[2] N. Minh, F. Liu, P. Staszak, T. Stillwagon, and J. Van Ackeren, "Monolithic Solid Oxide Fuel Cell Fabrication Development," *1988 Fuel Cell Seminar*, Long Beach, CA, p. 108, October 23 to 26, 1988.

[3] T. Kenjo and M. Nishiya, *Solid State Ionics*, 57, 295 (1992)

PULSED LASER DEPOSITION AND DC-SPUTTERING OF YTTRIA STABILISED ZIRCONIA FOR SOLID OXIDE FUEL CELL APPLICATIONS

B. Hobein[1,2], W.A. Meulenberg[1], F. Tietz[1], D. Stöver[1], E.W. Kreutz[2], M. Cekada[3], P. Panjan[3]

[1]Forschungszentrum Jülich, Institute for Materials and Processes in Energy Systems I, 52425 Jülich, Germany
[2]Lehrstuhl für Lasertechnik (LLT), Rheinisch-Westfälische Technische Hochschule Aachen, Steinbachstr. 15, 52074 Aachen, Germany
[3]Institute Jozef Stefan, Department of Thin Films and Surfaces, Jamova 39, 1000 Ljubljana, Slovenia

ABSTRACT

Solid oxide fuel cells (SOFC) are developed for applications especially in stationary energy supply. Yttria stabilised zirconia (YSZ) films were grown by pulsed laser deposition (PLD) from a YSZ target and by reactive DC-sputtering from a metallic Zr-Y target. The films have been deposited under oxygen atmospheres on porous NiO/YSZ substrates at 200 to 700 °C. YSZ films were obtained in the range of 1 to 8 μm thickness. The films have been investigated with respect to surface morphology, microstructure and film-substrate interaction at the interface . The film morpholgy varied from porous columnar to dense granular microstructure depending on the oxygen pressure and the substrate temperature. In the oxygen pressure regime of 0.01 to 0.5 mbar and 0.65×10^{-4} to 6.0×10^{-4} mbar the films consisted of cubic YSZ for PLD and sputtering whereas at lower oxygen pressures metallic and tetragonal phases appeared, too.

INTRODUCTION

The electrolyte thickness in anode-supported SOFC is about 5 to 20 μm [1]. In order to reduce the operating temperature from 800 °C down to 650 °C, while maintaining the electrochemical power density, 8 mol% Y_2O_3 stabilised zirconia

(8YSZ) electrolyte films of 1 to 2 μm thickness have to be realised [2]. Due to the advantages such as being able to transfer the stoichiometry from the target onto the substrate, film to film reproducibility and single-phase purity, pulsed laser deposition (PLD) is an interesting tool to deposit ceramic thin films. Dense, crack free and also epitaxial YSZ films have been deposited successfully by PLD on various substrates like polymethylmethacrylate (PMMA), polycarbonate (PC), fused silica, Si(100) or sapphire [3,4,5]. Previous investigations have shown the successful deposition of YSZ films by DC magnetron sputtering on various substrates such as steel, silicon, silica glass, platinum and also porous NiO/YSZ cermets [6,7]. It was reported that dense and impervious YSZ films up to 16 μm thickness had been prepared as single-phase cubic phase structure. The porous NiO/YSZ substrate used was of relatively high porosity up to 32% [7]. In the present work pulsed laser deposition and DC-sputtering was applied to deposit thin 8YSZ films on dense (porosity < 2 %) NiO/YSZ surfaces for fuel cell applications.

EXPERIMENTAL

Substrate preparation

The planar NiO/YSZ anode substrates with a thickness of approx. 2 mm were manufactured by warm pressing [1]. A less porous anode functional layer (NiO/YSZ) of 10 μm thickness was deposited on the presintered NiO/YSZ anode substrate by vacuum slip casting and sintered at 1400 °C for 4 h. After sintering the porosity of the functional layer was less than 2%. The substrates were cut to parts of 10×10 mm^2 in size and polished on a smooth pad using diamont paste with a grain size of 3 μm.

PLD setup

The target used during PLD was pressed and sintered 8YSZ powder (Tosoh) and mounted onto a rotating holder in a vacuum chamber with background pressure of 1×10^{-5} mbar. The angle between target and substrate normal was 45 °. The distance between target and substrate was 30 mm. The substrate was mounted on a resistance heater ranging from room temperature to 600 °C with a heating and cooling rate of 15 K/min. A KrF excimer laser (248 nm) was used at a pulse energy of 220 mJ (fluence=2.8 J/cm²) and a repetition rate of 30 Hz. The incident angle between the laser beam and the target normal was 45 °. YSZ thin films were deposited at 400 to 600 °C on the sintered substrates using oxygen as processing gas with pressures ranging from 0.01 to 0.5 mbar.

DC-sputtering setup

During deposition by DC reactive sputtering (Sputron, Balzers AG) a metallic ZrY target (80:20 at.%) was used. The background pressure in the vacuum chamber was below 1×10^{-6} mbar, while the pressure of argon during deposition was 2.0×10^{-3} mbar. Substrates of 50×50 mm^2 and 25×25 mm^2 in size were used and heated to 700 °C by halogen lamps. For comparison additional coatings were prepared on polished alumina ceramics. YSZ films were deposited at 500 °C to 700 °C using oxygen as the reactive gas with partial pressures ranging from 0.65×10^{-4} to 6.0×10^{-4} mbar. With 60 W/cm^2 DC power on the target a deposition rate of 26 nm/min was achieved for a substrate to target distance of about 200 mm.

Film characterisation

The phase compositions of the films were analysed by X-ray diffraction (XRD) using Cu Kα radiation. Scanning electron microscope (SEM) pictures and energy dispersive X-ray spectroscopy (EDS) were used to determine the thickness, the morphology and the general composition of the deposited films. A profilometer (Rank Taylor Hobson/ Talysurf Series) was used to measure the surface roughness of coated and uncoated substrates. The film thickness was measured with an appropriate software tool (analySIS by Soft Imaging System).

RESULTS AND DISCUSSION

Film thickness and roughness

The thickness of the PLD films on NiO/YSZ substrates varied due to the different orientations of the ablated target particles. The film thickness varied from the margins to the centre of the substrate (plasma) from 1.5 μm to 2.5 μm. At the chosen deposition parameters (T_{sub}= 500 °C, $p(O_2)$= 0.05 mbar) the growth rate at the centre of the plasma plume was 0.083 nm/pulse.
The roughness of the coated film surface was in the same range as the roughness of the substrate.

In the case of the DC-sputtered YSZ films no significant variations in the film thickness could be measured across each sample size.

Film morphology and phase composition

In the case of the YSZ films deposited by PLD at 600 °C with oxygen pressures ranging from 0.01 mbar to 0.5 mbar on NiO/YSZ substrates the X-ray diffraction patterns showed in all cases the YSZ cubic (c-YSZ) phase. Reducing the deposition temperature down to 500 °C amounts of the tetragonal (t-YSZ) YSZ phase at low oxygen pressures were measured. At 400 °C deposition temperature and oxygen pressures under 0.2 mbar amounts of the rhombohedral $Zr_3Y_4O_{12}$ phase [8] were detected (Fig. 1). At oxygen pressures lower than 0.05 mbar and deposition temperatures between 400 and 600 °C a partial reduction of NiO grains to Ni at the substrate film interface was observed (Fig. 1c). The deposited films were dense and crack free at oxygen pressures below 0.5 mbar. At higher pressures the morphology of the films changes to the columnar structure. The columns grew in the direction of the target normal.

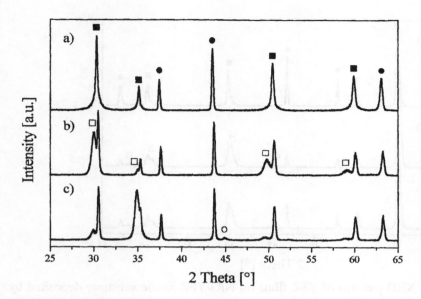

Fig. 1 XRD patterns of YSZ films deposited by PLD on NiO/YSZ anode substrate at 400 °C and different oxygen pressures: a) 0.5 mbar, b) 0.1 mbar and c) 0.01 mbar. The symbols shown correspond to c-YSZ (■), $Zr_3Y_4O_{12}$ (□), Ni (O), NiO (●).

DC-sputtered XRD-patterns of YSZ films were measured on NiO/YSZ anode substrates and alumina substrates. The YSZ films were deposited at 500 °C, 600 °C and 700 °C with partial pressures of oxygen ranging from 0.65×10^{-4} mbar to 6.0×10^{-4} mbar. In all cases, but much more pronounced at low partial pressures of oxygen, the diffraction pattern showed besides c-YSZ also significant amounts of t-YSZ as shown in Fig. 2. At oxygen partial pressures of 2.3×10^{-4} mbar and below reflections of metallic Ni were detected (Fig. 2). In all cases dense and crack-free films were deposited. SEM cross-section micrographs are shown in Fig. 4. After postannealing of the coated substrates at 1100 °C for 3 h in air, the tetragonal YSZ phase was completely transformed to the cubic YSZ phase.

The diffraction patterns of the films coated on polished alumina substrates showed c-YSZ at oxygen pressures of 2.3×10^{-4} mbar and above (Fig. 3). At lower oxygen pressures down to 0.65×10^{-4} mbar wider reflections of the YSZ pattern were measured according to smaller crystallites and/or increased disorder. Even an amorphous film was deposited as shown in Fig. 3c. Without any oxygen supply the metallic phase of zirconium was deposited.

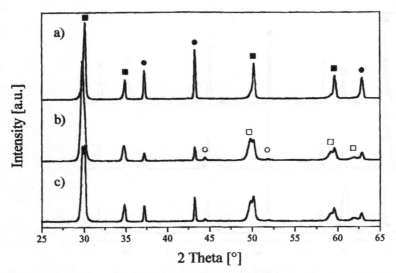

Fig. 2   XRD patterns of YSZ films on NiO/YSZ anode substrate deposited by DC-sputtering at different oxygen partial pressures and substrate temperatures: a) 6.0×10^{-4} mbar at 500 °C, b) 2.3×10^{-4} mbar at 700 °C and c) 0.65×10^{-4} mbar at 600 °C. The symbols shown correspond to c-YSZ (■), t-YSZ (□), Ni (O), NiO (●).

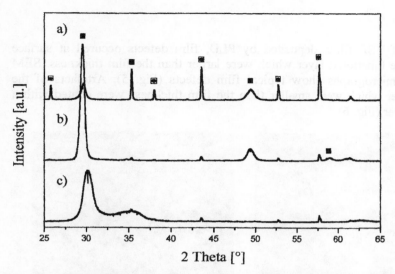

Fig. 3 XRD patterns of YSZ films on alumina substrates deposited by DC-sputtering at different oxygen partial pressures and substrate temperatures: a) 6.0×10^{-4} mbar at 500 °C, b) 2.3×10^{-4} mbar at 700 °C and c) 0.65×10^{-4} mbar at 600 °C. The symbols shown correspond to c-YSZ (■) and alumina (◙).

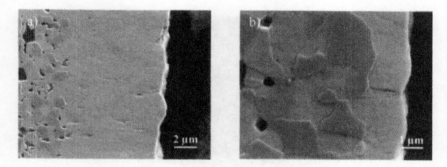

Fig. 4 SEM cross section micrographs of YSZ films on NiO/YSZ substrates deposited by DC-sputtering with oxygen pressures and substrate temperatures of (a) 0.65×10^{-4} mbar at 600 °C and (b) 6.0×10^{-4} mbar at 500 °C.

Film defects

In the case of YSZ films deposited by PLD, film defects occured at surface artefacts of the functional layer which were larger than the film thickness. SEM cross-section micrographs show typical film defects (Fig. 5). Artefacts of the functional layer which were smaller than the film thickness were coated with a dense YSZ layer (Fig. 6).

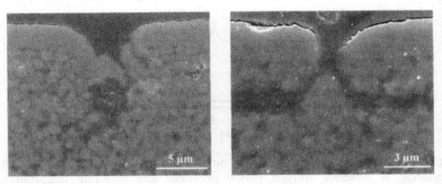

Fig. 5   SEM cross-section micrographs of YSZ film defects deposited by PLD on NiO/YSZ substrates.

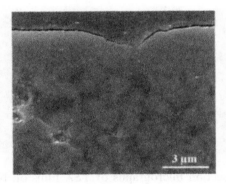

Fig. 6   SEM cross-section micrograph of YSZ film deposited by PLD on NiO/YSZ substrate artefact smaller than the film thickness.

CONCLUSION

Thin YSZ films with a thickness of 1 to 2.5 μm have been obtained by pulsed laser deposition on nearly dense (porosity < 2 %) NiO/YSZ anode substrates at 400 to 600 °C for fuel cell applications. The film thickness on the substrate was strongly depending on the geometry of the plasma plume. Relative film thickness variations of 60 % were measured on substrates of 10×10 mm^2 in size. Low oxygen pressures (< 0.2 mbar) and deposition temperatures ($\leq$ 500 °C) during deposition induced dense YSZ films grown in cubic, rhombohedral and tetragonal modification. At higher oxygen pressures cubic YSZ films were deposited in columnar structure. Film defects occured at surface artefacts of the functional layer which were larger than the film thickness.

In the case of DC-sputtering 2 to 8 μm thick YSZ films have been deposited on NiO/YSZ and alumina substrates. No significant inhomogeneities in film thickness on the substrates were measured. In all cases dense and crack free YSZ coatings in cubic structure were deposited. Low oxygen partial pressures below 2.3×10^{-4} mbar led to amounts of tetragonal YSZ in the deposited films.

ACKNOWLEDGEMENTS

Financial support from the Deutsche Forschungsgemeinschaft (DFG) under contracts KR 516/22 and STO 382/1 as well as from the German Ministry of Education and Research (BMBF) and the Slovenian Ministry for Science and Technology (MZT) under contract no. SVN 99/034 is gratefully acknowledged.

REFERENCES

[1] D. Simwonis, H. Thülen, F.J. Dias, A. Naoumidis, D. Stöver, J. Mat. Proc. Technol. 92-93 (1999) 107

[2] B. C. H. Steele, Proc. 1st Eur. Solid Oxid Fuel Cell Forum, ed. U. Bossel, Vol.1, (1994), 375

[3] J. Gottmann, E.W. Kreutz, Surf. Coat. Technol. 116-119 (1999) 1189

[4] J.C. Delgado, F. Sanchez, R. Aguiar, Y. Maniette, C. Ferrater, M. Varela, Appl. Phys. Lett. 68 No. 8 (1996) 1048

[5] G.A. Smith, Li-Chyong Chen, Mei-Chen Chuang, Mat. Res. Soc. Symp. Proc. 235 (1992) 429

[6] A. F. Jankowski and J. P. Hayes, Surface and Coatings Technology, 76-77, Part 1, (1995), 126

[7]P.K. Srivastava, T. Quach, Y.Y. Duan, R. Donelson, S.P. Jiang, F.T. Ciacchi, S.P.S. Badwal, Solid State Ionics 99 (1997) 311
[8]H. G. Scott, Acta Cryst. B33 (1977) 281-282

MICROSTRUCTURE – ELECTRICAL PROPERTY RELATIONSHIP IN NANOCRYSTALLINE CeO_2 THIN FILMS

V. Petrovsky, B. P. Gorman[*], H. U. Anderson and T.Petrovsky
Electronic Materials Applied Research Center, University of Missouri–Rolla, Rolla, MO 65409

ABSTRACT

Optical and electrical properties of nanocrystalline cerium oxide thin films undoped and doped by Gd and Sm were investigated. Two routes (colloidal suspensions and polymeric precursors) were used for film preparation using silica and alumina as substrates. The combination of the two preparation techniques made possible to obtain films with grain size from 5 to 100nm and wide region of the porosity (from less than 5% to 50%). The values of thickness, density and grain size were obtained from the optical measurements. Nanocrystalline films with very small grain size (less than 10nm) show small difference in the value of conductivity and activation energy for all film composition and preparation techniques investigated. For ceria films deposited on silica substrates, it was found that the conductivity decreases and activation energy increases for annealing temperatures as low as 600°C. In contrast, films that were deposited on sapphire substrates showed conductivity increases and activation energy decreases as the annealing temperature and grain size increases and approached the values obtained for dense polycrystalline specimens ($\sigma \sim 0.01$ S/cm at 600°C, $E \sim 0.7$ eV) at the annealing temperature of 1000°C (grain size ~ 60 nm).

INTRODUCTION

CeO_2 (ceria) based materials have been widely investigated, mainly due to high ionic conductivity. This property makes them highly desirable as electrolytes and electrodes in solid oxide fuel cells and oxygen separation membranes. Recent

---

[*] present address: Dept. of Materials Science, University of North Texas, Denton TX 76203

reviews of ceria-based materials for these applications give highly detailed properties of this system [1-3].

Ceria has a fluorite structure over a wide temperature and oxygen partial pressure range. Ce^{4+} reduces to Ce^{3+} with a decrease in oxygen partial pressure, which greatly increases the electronic portion of the conductivity. The partial pressure at which this occurs is highly dependent upon the impurity level of acceptor dopants in the raw material [4,5]. For electrolyte applications where the electronic portion of the conductivity should be kept to a minimum, CeO_2 is typically doped with acceptors such as Gd^{3+} or Sm^{3+} [6-9].

Thin film undoped and doped CeO_2 have also been extensively studied [10-12]. These studies have shown that nanocrystalline films with high ionic conductivity can be prepared, however, there may be problems associated with impurity effects related to impurities contained either in the starting materials or those resulting from interactions between the films and the substrates. Studies have also shown that increased electronic conductivity occures in the nanocrystalline region due to decreased oxygen vacancy formation energy at these grain boundaries [10,13-17]. In an effort to understand the observed effects, this investigation has focused on the study of the electrical conductivity and optical properties in porous nanocrystalline CeO_2 (99.999% purity) thin films processed using a colloidal deposition technique and dense nanocrystalline films produced by a polymer precursor process (99% purity). The grain size and porosity were varied by the sintering, and the conductivity was measured using DC 2-probe and 4-probe technique.

EXPERIMENTAL

The high purity nanocrystalline powders which were used for colloidal deposition were prepared by an aqueous precipitation processing route using 99.999% cerium (III) nitrate (Alfa Aesar, Inc.) which was dissolved in distilled water and precipitated to cerium hydroxide particles using hydrogen peroxide and ammonium hydroxide, as illustrated by Djuricic and Pickering [18]. The hydrated powder was then annealed at 150°C in order to form the full fluorite structured CeO_2 powder. Samarium doped nanocrystalline ceria was obtained from NexTech Materials, Inc. for preparation of acceptor doped porous films.

Aqueous colloidal suspensions were prepared by mixing 10 weight percent of the CeO_2 powder in pH 5 distilled water (balanced with HNO_3) along with 5 weight percent butoxyethanol to aid in drying and decrease the wetting angle. The butoxyethanol also partially polymerizes due to the acidic content of the solution, and as a consequence aids in increasing the solution viscosity and in the formation of a continuous film. The powder was then dispersed in the solution using a high intensity ultrasonic probe for about one hour and than filtered through a 0.45μm glass fiber filter (Whatman, Inc.) in order to remove any foreign matter.

For the polymeric precursor preparation cerium and gadolinium nitrates (99% purity) were used as the cation sources and ethylene glycol was polymerizing agent. The nitrates were mixed in the desired ratio and dissolved in the water-ethylene glycol mixture. Partial polymerization was accomplished by heating the solution at 70°C for 75h. Polymerized precursors were also filtered through Whatman, Inc. glass fiber filters. Details of the polymeric solution preparation and dense film deposition are presented elsewhere [10].

Thin films were prepared by spin coating the precursor on quartz and sapphire substrates. Optical quality, both sides polished (0001) oriented sapphire substrates were used for both the optical measurements as well as the DC conductivity measurements. Single crystal quartz substrates were used only for electrical measurements to investigate possible influence of the substrate material on electrical properties of the film.

The films were deposited on the substrates by spinning at 1500rpm for 30 seconds and subsequently dried at 70°C for an hour. The films were then heat treated at 350°C for an hour and cooled back to room temperature after which subsequent coatings could be deposited. Sintering of the films was done in a small box furnace using a ramp rate of 5°C/min to the maximum temperature, with a 2 hour hold at maximum temperature. The films were characterized according to their grain size, thickness and density using X-ray diffraction, mechanical profilometry, ellipsometry, and UV-Vis spectroscopy, as shown in previous studies [11,12]. The resulting film density and thickness are shown in Fig.1 for colloidal films. It can be seen that after 400°C the initial films have density of approximately 50% and grain size of 5 nm after 400°C and densify to 85% with a grain size of 60 nm after annealing at 1000°C.

X-ray diffraction showed that the films are single-phase fluorite structured and have no reaction with the substrate visible in the XRD spectra over the investigated temperatures. Peak broadening grain size calculations (Reitfeld analyses) from the diffraction patterns correlate well with those in FESEM images (Fig.2).

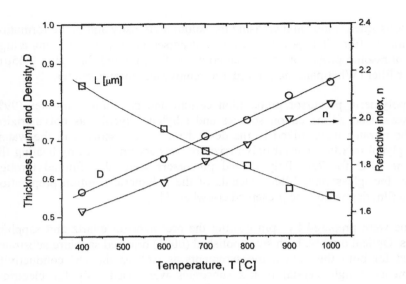

Figure 1. Change in film thickness, refractive index and density as a function of annealing temperature for colloidal processed CeO_2 thin films [11].

Figure 2. FESEM images of CeO_2 thin films on sapphire substrates after annealing at 700°C for 2 hours. (a) - the surface (b) - fracture cross section

Materials for Electrochemical Energy Conversion and Storage

DC electrical conductivity measurements were done in 4-probe and 2-probe configuration. Silver paste was used as the 4 electrode contacts for each sample. The contacts were painted on and annealed at 600°C for 1 hour (with the exception of samples sintered at 400°C, where the electrodes were annealed at the same temperature for 1 hour). After annealing, the electrodes were attached to Pt wire leads, which ultimately were contacted to the testing apparatus.

A Keithley 4915A electrometer connected to a personal computer was used as a measuring device. It was possible to make current - voltage measurements or pulsing technique in both 2-probe and 4-probe configurations. The current - voltage characteristics were used to check the ohmic character of the contacts. Delay time between each voltage application was varied in order to minimize hysteresis effects connected with the high resistance of the samples. I-V characteristics were linear, if a sufficiently long delay time was chosen. The magnitude of the delay depended on the temperature and increased up to 1s for the lower temperature region.

Pulsing techniques using direct current are particularly useful for measurements of the high resistance samples (Fig.3). With this technique it is possible to exclude the relaxation process, provide accurate measurements on the plateau and check the symmetry of the sample properties by comparing the measurements at the positive and negative bias. This technique was used for 2-probe and 4-probe measurements of the sample resistance.

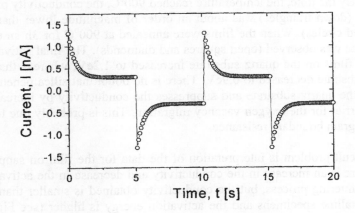

Figure 3. Pulsing technique for the measurements of ceria films.

RESULTS

Two basic features differ nanocrystalline films from microcrystalline dense specimens. First of all the starting material has extremely small grain size (almost amorphous), and it remains small (< 60 nm) over all of the temperature regions of this investigation. This high degree of disorder can effect electrical properties of the material. The question is how big this effect could be and what is the critical size of the grains to have the conductivity comparable with microcrystalline specimens of the material?

The second difference is the presence of the substrate which can effect the properties of the film by reaction with the film. The possible effect of the substrate was investigated by deposition the dense nanocrystalline films on quartz and sapphire substrates under the same conditions by the polymeric precursor process. The results of this investigation are shown in Fig.4 as an Arrhenius plot of the electrical conductivity for dense gadolinium doped ceria (CGO) films.

It can be seen from the figure that for the films annealed at 600°C or lower, the conductivity does not depend on the substrate and the activation energy is the same ~1.1eV until the film has reached temperatures exceeding about 700°C (open circles and up triangles). When the measurement and/or annealing temperature exceeds temperatures of about 700°C, the slope for the film on sapphire substrate remains the same up to 900°C while for the film on the quartz substrate shows a decrease in slope. These changes in the conductivity are nonreversible. By the time, the temperature reached 900°C, the conductivity of the film on quartz (down triangles) was about an order of magnitude lower than on sapphire (closed circles). When the films were annealed at 900°C for 3h an even larger difference was observed (open squares and diamonds). The final activation energy for the films on the quartz substrate increased to 1.2eV, whereas that on the sapphire substrate decreased to 0.9eV. There is no doubt that silica penetrates the film from the quartz substrate and suppresses the conductivity by increasing the hopping barrier for the oxygen vacancy migration. This is probably due to an increase in the grain boundary resistance.

The more difficult problem is interpretation of the data for the film on sapphire substrates. There is an increase in the conductivity and decrease in the activation energy at the sintering process, but the conductivity obtained is smaller than for dense polycrystalline specimens and the activation energy is higher (see Fig.4). Different explanations of this fact are possible. The first explanation is with the small grain size the grain boundary resistance is dominating the total resistance, the second explanation is the presence of uncontrolled contaminations in the film and the third is diffusion of the aluminum into the film from the sapphire substrate.

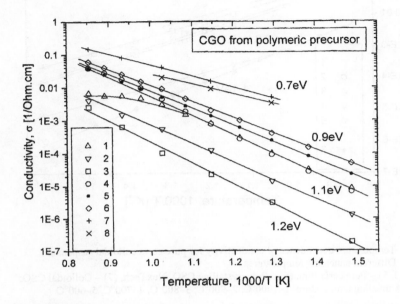

Figure 4. Temperature dependence of the conductivity for gadolinium doped ceria films deposited on quartz and sapphire substrates.
(1-3) – Film on quartz substrate, (4-6) – Film on sapphire substrate,
(7,8) – Polycrystalline specimens – literature data (7- CSO, NexTech; 8-CGO, Stele [6)]);
(1,4) – Increase temperature, (2,5) – Decrease temperature, (3,6) – Annealed at 900°C for 3 h.

The way to differentiate these effects is to use another precursors, prepared by different technology. Colloidal suspensions of samarium doped and extremely pure ceria were used for this purpose (Fig.5). It is seen from the figure that again the initial conductivity and the energy activation (E~1.1eV) are the same for the films prepared from both precursors (solid circles and diamonds). This result is quit similar to that observed for the dense films obtained from ceria-gadolinium polymeric precursor: activation energy is the same and the conductivity is only slightly smaller (which can be connected with high porosity of the colloidal films). It looks like initial conductivity is controlled by extremely small grain size and is not sensitive to the composition or precursor nature.

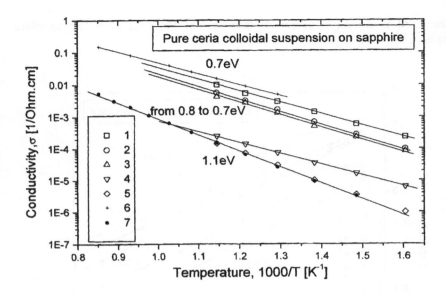

Figure 5. Temperature dependence of conductivity for pure ceria colloidal suspension.
Different annealing temperatures.
(1-5) – Pure ceria film, (6) – Polycrystalline CSO, NexTech, (7) – Colloidal CSO;
Annealing temperature: 1–1000°C, 2–900°C, 3–800°C, 4–700°C, 5–600°C.

The properties of the colloidal samarium doped ceria prepared film did not change with annealing up to1000°C (temperature dependence of the conductivity remains the same), which possibly may be connected with impurities in the initial powder. (This is purely speculation, since the impurity levels have not be measured).

In the case of the films prepared with the undoped, higher purity colloidal ceria very different results were observed. As can be seen from Fig.5 the conductivity starts to increase after 700°C sintering and the activation energy decreases. The sintering at 1000°C gives the conductivity and activation energy comparable with that reported for polycrystalline specimens (see Fig.5).

The dependences of the conductivity and activation energy are presented on Fig.6 as the function of grain size. It can be seen that the activation energy decreases quickly as grain size increases from 10 to 20nm and than decreases very slowly from 0.8 to 0.7eV for further grain size increase. That suggests that the barrier for the motion of oxygen vacancies changes little after grain size reaches about 20nm and is controlled by the grains rather than grain boundaries. The question that needs to be addressed is why the conductivity of undoped film increases to the level observed for dense polycrystalline samarium doped specimens? A possible explanation would be interdiffusion of the sapphire substrate with the film. If this were the case, the Al_2O_3 can play role of an acceptor impurity (similar to Sm_2O_3 or Gd_2O_3). Acceptor doping should be not effective for stabilized materials (Sm or Gd doped), because the addition of acceptor impurity beyond the level already there will do little to increase the conductivity. However, in the case of undoped material, such acceptor doping could result in substantial increase in the conductivity, proportionally to the acceptor concentration without changing the activation energy. This behavior is in full agreement with the experimental data presented. It is necessary to emphasize that at this time there are no data in this study to substantiate the hypothesis that such diffusion occurred, however, the acceptor behavior of alumina in the ceria-alumina-silica solid solution was shown earlier [19]. More precisely, it was shown that the ratio of Ce^{3+} / Ce^{4+} increases with the increase of alumina content.

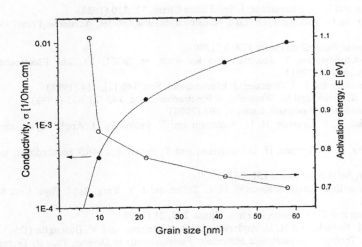

Figure 6. Electrical conductivity at 600°C and activation energy as a function of grain size. Pure ceria colloidal films.

CONCLUSIONS

- The interaction of ceria films with silica limits the use of silica as a substrate to temperatures below 600-700°C.
- Grain size limits the conductivity only in the very small grain size region (less than 10-20nm). The conductivity and activation energy is not sensitive to the composition of the film or precursor used in this grain size region.
- It is possible to achieve small activation energy (~0.7eV) and the conductivity comparable with the dense polycrystalline specimens (0.01S/cm at 600°C), if pure starting materials are used for the film preparation.

REFERENCES

1. M. Mogensen, N. M Sammes and G. A. Tompsett, Solid State Ionics **129**, 63 (2000).
2. H. Inaba and H. Tagawa, Solid State Ionics **83**, 1 (1996).
3. B. C. H. Steele, in: High Conductivity Solid Ionic Conductors, ed. T. Takahashi, (World Scientific, Singapore, 1989).
4. E. K. Chang and R. N. Blumenthal, J. Solid State Chem. **72**, 330 (1988).
5. H. L. Tuller, in: O. Toft Sorensen (Ed.), Nonstoichiometric Oxides, Academic Press, 1981, p. 271.
6. B. C. H. Steele, Solid State Ionics **129**, 95 (2000).
7. B. Zachau-Christiansen, T. Jacobsen and K. West, in: SOFC III, The Electrochemical Society, Inc., p. 104 (1991).
8. M. Godickemeier and L. J. Gauckler, J. Electrochem. Soc. **145** [2], 414 (1998).
9. H. Uchida, H. Suzuki and M. Watanabe, J. Electrochem. Soc. **145** [2], 615 (1998).
10. I. Kosacki and H. U. Anderson, Ionics **6**, 294 (2000).
11. V. Petrovsky, B. P. Gorman, H. U. Anderson and T. Petrovsky, J. Appl. Phys., submitted (2001).
12. V. Petrovsky, B. P. Gorman, H. U. Anderson and T. Petrovsky, MRS Proceedings, in print (2000).
13. H. L. Tuller, Solid State Ionics **131**, 143 (2000).
14. Y.-M. Chiang, E. B. Lavik, I. Kosacki, H. L. Tuller and J. Y. Ying, Appl. Phys. Lett. **69** [2], 185 (1996).
15. J. –H. Hwang and T. O. Mason, Z. Phys. Chem. **207**, 21 (1998).
16. T. Suzuki, I. Kosacki and H. U. Anderson, in: P. Vincenzini and V. Buscaglia (Eds.) *Mass and Charge Transport in Inorganic Materials: Fundamentals to Devices, Part A*, Techna Srl, 2000, p. 419-29.
17. A. Atkinsson and C. Monty, in: L. C. Dufour, et al. (Ed.), Surfaces and Interfaces of Ceramic Materials, Kluwer Academic, Dodrecht, The Netherlands, 1989, p. 273.
18. B. Djuricic and S. Pickering, J. Eur. Cer. Soc. **19**, 1925 (1999).
19. Sin-Lung Lin, Chii-Shyang Hvang and Jyh-Fu Lee, J.Mater.Res, **11**, 2641 (1996).

ELECTRICAL MEASUREMENTS IN DOPED ZIRCONIA-CERIA CERAMICS

Cesar R. Foschini, Leinig Perazolli
José A. Varela
Instituto de Química-UNESP
Araraquara, São Paulo, Brazil
14801-970

Dulcina P.F. Souza,
Pedro I. Paulin Filho
Dpt. de Eng. de Materiais - UFSCar
S. Carlos, SP, Brasil
13565-905

ABSTRACT

Zirconia-ceria powders with 12 mol % of CeO_2 doped with 0.3 mol% of iron, copper, manganese and nickel oxides were synthesized by the conventional mixed oxide method. These systems were investigated with regard to the sinterability and electrical properties. Sintering was studied considering the shrinkage rate, densification, grain size, and phase evolution. Small amount of dopant such as iron reduces sintering temperature by over 150°C and more than 98% of tetragonal phase was retained at room temperature in samples sintered at 1450°C against 1600°C to stabilize the tetragonal phase on pure ZrO_2-CeO_2 system. The electrical conductivity was measured using impedance spectroscopy and the results were reported. The activation energy values calculated from the Arrhenius's plots in the temperature range of 350-700°C for intragrain conductivities are 1.04 eV.

INTRODUCTION

The zirconium oxide, known as zirconia, has three polymorphic phases:

$$\text{Monoclinic} \xleftarrow{1170°C} \text{Tetragonal} \xleftarrow{2370°C} \text{Cubic} \xleftarrow{2680°C} \text{Liquid}$$

The preparation of tetragonal zirconia can be made by either adding 3 mol% Y_2O_3 [1] or 12 mol% CeO_2 [2]. The ceria-zirconia system is being developed as an alternative to the ytria-zirconia system because of its better performance in a moisture environment [3], good mechanical properties [4], lower price and wider range of solubility [5]. Unfortunately, cerium ions are less effective than yttrium ones in stabilizing the tetragonal phase of zirconia, and generates a low sinterability system [6]. In the case of yttria-zirconia, the tetragonal phase is retained due to the presence of oxygen vacancies, which account for the charge neutrality. In the case of ceria-zirconia, the Ce ions also substitute for Zr^{4+} ions, but they can have two valence states: Ce^{3+} and Ce^{4+}[7].

The effect of sintering aids such as copper and manganese ions on the sintering behavior of ceria-stabilized zirconia was investigated by S. Mashio et al[8]. The authors verified that MnO_2 as well as CuO were very effective in improving the tetragonal phase content retained at room temperature.

In this work we have studied the effect of the sintering aids such as nickel, iron, copper and manganese on the electrical properties of the ceria-zirconia system. With the use of these sintering aids for Ce-doped tetragonal zirconia polycrystals (Ce-TZP), we were able to produce a dense ceramic material, with an uniform microstructure, stabilized more than 98% in the tetragonal phase and sintered at temperatures up to 150°C lower than that for undoped CeO_2-ZrO_2 system.

EXPERIMENTAL PROCEDURE

Commercial zirconium oxide (>99.9% in purity) and cerium oxide (>99.0% in purity) were mixed using zirconia balls in isopropyl alcohol medium to produce powders with the following composition 88 mol% ZrO_2-12 mol% CeO_2. In order to study the influence of additives, the dried ZrO_2-12%CeO_2 powder was divided into five batches: (1) as prepared powder labeled ZrCe; (2) ZrCe + 0.3 mol% NiO, labeled ZrCeNi; (3) ZrCe + 0.3 mol% Fe_2O_3, labeled ZrCeFe; (4) ZrCe + 0.3 mol% CuO, labeled ZrCeCu; (5) ZrCe + 0.3 mol% MnO, labeled ZrCeMn. Nickel and iron nitrates, copper and manganese acetates (Riedel del Haën) were added and ball milled in plastic jars with zirconia balls for 30 h with an organic defloculant. Powders were calcinated at 500°C to eliminate volatiles, ball milled in isopropyl alcohol with the addition of 1 wt% of polyvinyl butyral (PVB), dried at room temperature, and sieved through an 80 mesh nylon sieve. Pellets of 12 mm diameter were isostatically pressed at 270 MPa and sintered in the range of 1400 to 1600°C for 1h in air.

The apparent densities of the sintered pellets were measured using Archimedes' method with distilled water. XRD (Siemens D-5000) was used to analyze the sintered pellets for phase identification. SEM (Topcon SM 300) was used to study the microstructure. Samples were ground with successive grades of SiC papers; polished with 15, 6, 3 and 1 μm grits of diamond paste and thermal etched at 50 degrees below the sintering temperature during 10 min. The dilatometric study was performed in a dilatometer (Netzsch 402E) up to 1500°C using a constant heating rate of 10°C/min. Electrical measurements were performed on polished pellets with platinum electrodes applied by painting (Demetron 308A). The grain electrical conductivity was measured by impedance spectroscopy technique using the impedance analyzer (HP 4192A LF) in the frequency and temperature range of 5Hz to 13MHz and 300 to 700°C respectively, during heating with a hold time of 30 min in each temperature.

RESULTS AND DISCUSSION

Table I summarizes the values of sintering temperature, apparent density and percentage of tetragonal phase retained at room temperature after sintering. The sintering temperature was chosen based on the X-ray diffraction.

Table I. Sintering temperature, apparent density, percentage of tetragonal phase retained at room temperature and average grain size of ZrO_2-CeO_2 samples undoped and Ni, Fe, Cu and Mn doped.

Sample	Sint. Temp. (°C)	Density (g/cm^3)	Percentage of tetragonal phase (%)	Average grain size (μm)
ZrCe	1600	5.65	99.4	1.22±0.58
ZrCeNi	1500	5.79	99.7	1.27±0.59
ZrCeFe	1450	5.63	98.6	1.19±0.59
ZrCeCu	1500	5.87	100.0	1.21±0.58
ZrCeMn	1500	5.76	99.7	1.18±0.57

Figure 1 shows XRD patterns of samples undoped and Fe_2O_3 doped sintered at different temperatures. As one can observe, the presence of additives can decrease the stabilization temperature of tetragonal phase up to ~150°C in the case of iron.

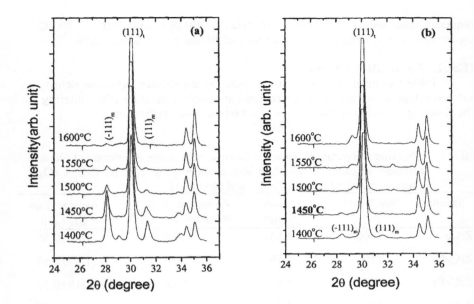

Figure 1: X-ray diffraction patterns of (a) ZrO_2-12mol%CeO_2 and (b) ZrO_2-12mol%CeO_2 + 0.3 mol% Fe_2O_3 sintered in a temperature range of 1400 to 1600°C.

The dilatometry data also indicated that the additives promote the ZrCe densification and that copper and iron are the most efficient. From Figure 2, we can observe that the starting temperature of shrinkage was 1220°C for ZrCe sample and 1100°C for compositions with iron and copper. Dense samples were obtained at 1425°C for the ZrCe and at 1305°C for ZrCeFe composition.

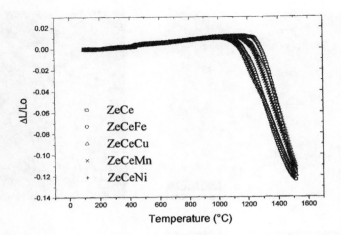

Figure 2: Dilatometric curves for pure ZrO_2-CeO_2 and Fe_2O_3, CuO, MnO and NiO doped samples.

Fig. 3(a) and (b) shows microstructures of specimens ZrO_2-CeO_2 and ZrO_2-CeO_2-Fe_2O_3. The results of grain size measurements are given in Table I. The average grain size, measured by image analysis, was similar for all specimens examined and was in the range of 1.18 to 1.27 μm. It was also observed an inhomogeneous grain size distribution and practically no residual porosity.

Figure 3: Scanning electron micrographs of a) ZrCe; b) ZrCeFe; c) ZrCeMn; d) ZrCeCu; e) ZrCeNi.

Figure 4(a) and (b) show the impedance diagrams obtained at 397 and 598°C respectively. Measurements were performed in air. Through this technique it was possible to separate the contribution of the grain and the grain boundary resistivities. In Fig. 4(a) and (b), for each specimen, the arc on the left (next to the Y or the imaginary impedance axis) is due to the intragrain resistivity and the arc on the right (next to the intragrain resistivity arc) is due to the intergrain resistivity. The difference of intercepts of each arc on the X or real impedance axis gives resistivity associated with the intra or intergrain process. The arc on the extreme right (just appearing on the low frequency end of the impedance diagrams) is due to the relaxation of the electrode process. It can be observed that the presence of transition metals decreased the sample electrical resistivity of grain and grain boundary when compared with the composition ZrCe. The ZrCeFe composition showed the lowest resistance. This reduction promoted by the iron in the electrical resistance could be associated with the contribution of the electronic conduction due to the presence of Fe^{+3} and Fe^{+2}. However it can be observed that there is no significant alteration in the impedance spectra for the different compositions. That is, the spectra for all compositions are formed by simple semicircles with their centers on the axis x.

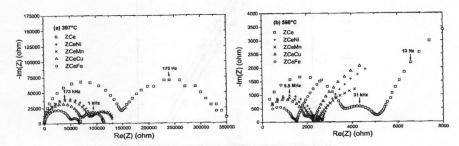

Figure 4: Impedance plots at (a) 397°C and (b) 598°C.

It was also observed in the Arrhenius's plots (Fig. 5) no variation in the slope for intragrain conductivities over the entire temperature range of measurements for the specimens with iron content. The activation energy values calculated from the Arrhenius's plots in the temperature range of 350-700°C for intragrain conductivities are 1.04 ± 0.02 eV. These results are in good agreement with the literature [9] and suggest that the conduction is mainly due to ionic diffusion. At high temperature range, the contribution from the grain boundary resistivity to the total resistivity is expected to be relatively small because the activation energy associated with oxygen-ion migration within the zirconia lattice is much lower than that for the grain boundary resistivity. In this way, the role of the grain boundary or the impurity phases (if present) becomes more critical in the low temperature range. The activation energy in the low temperature range consists of activation energy of defect migration as well as activation energy for dissociation of defect-associates whereas in the high temperature range where most defect-associates dissociate, the activation energy is likely to be mainly due to migration of oxygen-ion vacancies [10, 11].

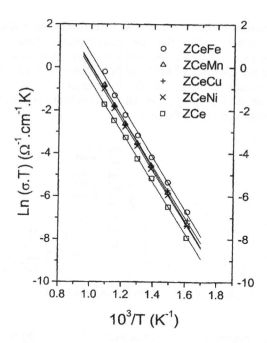

Figure 5: Arrhenius plots for intragrain conductivities measured in the temperature range of 350-700°C.

CONCLUSION

Small amount of dopants such as Ni, Fe, Cu and Mn transition metals reduces sintering temperature by over 150°C when compared with undoped ZrO_2-CeO_2 system.

The transition metals do not affect the conduction mechanism, as verified through the Arrhenius's plots, but increase the lattice conductivity up to three times in case of iron. The activation energy values for intragrain conductivities calculated from the Arrhenius's plots in the temperature range of 350-700°C are 1.04 eV.

ACKNOWLEDGEMENTS

This research was supported by FAPESP-Brazil, Process n° 98/13678-3.

REFERENCES

[1] L. Ruiz, and M. J. Readey, "Effect of Heat Treatment on Grain Size, Phase Assemblage, and Mechanical Properties of 3mol% Y-TZP," *J. Am. Ceram. Soc.*, **79** [9], 2331-2340 (1996).

[2] J. Wang, X. H. Zheng and R. Stevens, "Fabrication and microstructure-mechanical property relationships in Ce-TZPs," *J. Mater. Sci.*, **27** [19], 5348-5356 (1992).

[3] R. L. K. Matsumoto, "Aging Behavior of Ce-Stabilized Tetragonal Zirconia Polycrystals," *J. Amer. Ceram. Soc.*, **71** [3], C128-129 (1978).

[4] K. Sukuma and M. Shimada, "Strength, Fracture Toughness and Vickers Hardness of CeO_2-Stabilized Tetragonal ZrO_2 Polycrystals (Ce-TZP)," *J. Mater. Sci.*, **20**, 1178-1184 (1985).

[5] J. Wang, X. H. Zheng and R. Stevens, "Fabrication and Microstructure-Mechanical Property Relationships in Ce-TZPs," *J. Mater. Sci.*, **27**, 5348-53 (1992).

[6] H. Y. Zhu, "$CeO_{1.5}$ - Stabilized Tetragonal ZrO_2," *J. Mater. Sci.*, **29**, 4351-4356 (1994).

[7] P. Li and I. W. Chen, "Effect of Dopants on Zirconia Stabilization-An X-ray Absorption study: II, Tetravalent Dopants," *J. Am. Ceram. Soc.*, **77** [5], 1281-1288 (1994).

[8] S. Maschio, O. Sbaizero, S. Meriani and E. Bischoff, "Sintering Aids for Ceria-Zirconia Alloys," *J. Mater. Sci.*, **27**, 2734-2738 (1992).

[9] R. F. Reidy and G. Simkovich, "Electrical Conductivity and Point Defect Behavior in Ceria-Stabilized Zirconia," *Solid State Ion.*, **62** 85-97 (1993).

[10] B. C. H. Steele, in: *High Conductivity Solid Ionic Conductors, Recent Trends and Applications,* pp. 402-446, Edited by T. Takahashi, World Scientific, Singapore, 1989.

[11] S. P. S. Badwal, in: *Materials Science and Technology*, A Comprehensive Treatment p. 567, Edited by R. W. Cahn, P. Haasen and E. J. Kramer, Vol 11, VCH, Weinheim, Germany, 1994.

[12] S. P. S. Badwal, J. Drennan and A. E. Hughes, in: *Science of Ceramic Interfaces,* pp. 227-284, Edited by J. Nowotny, Elsevier, Amsterdam, 1991.

[13] R.Gerhardt and A. S. Nowick, "Grain-Boundary Effect in Ceria Doped with Trivalent Cations: I, Electrical Measurements". *J. Am. Ceram. Soc.*, **69** [9], 641-646, (1986).

[14] M. J. Verkerk, B. J. Middlehuis and A. J. Burggraaf, *Solid State Ion.* **6**, 159 (1982).

[15] M. Kleitz, H. Bernard, E. Fernandez and E. Schouler, in: *Science and Technology of Zirconia, Advances in Ceramics*, Vol 3, pp. 310-336. Edited by A. H. Heuer and L. W. Hobbs, The Am. Ceram. Soc. Inc., Columbus, OH, 1981.

[16] S. Kobayashi, "X-ray microanalysis of the boundary phase in partially stabilized zirconia (PSZ)," *J. Mater. Sci. Lett.*, **4** [3], 268-270 (1985).

[17] B.V. Narasimha Rao and T.P. Schreiber, "Scanning Transmission Electron Microscope Analysis of Solute Partitioning in a Partially Stabilized Zirconia," *J. Am. Ceram. Soc.*, **65** [3], C44-45 (1982).

[18] R. Chaim, D. G. Brandon and A. H. Heuer, "A Diffusional Phase Transformation in ZrO_2-4wt% Y_2O_3 Induced by Surface Segregation," *Acta Metll.*, **34** [10], 1933-1939 (1986).

[19] G. S. A. M. Theunissen, A. J. A. Winnubst and A. J. Burggraaf, "Segregation aspects in the ZrO_2-Y_2O_3 ceramics system," *J. Mater. Sci. Lett.*, **8** [1], 55-57 (1989).

[20] A. E. Hughes and S. P. S. Badwal, "Impurity and yttrium segragation in yttria-tetragonal zirconia," *Solid State Ion.*, **46** [3-4], 265-274 (1991).

[21] S. P. S. Badwal, "Electrical Conductivity of Single Crystal and Polycrystalline Ytria-Stabilized Zirconia," *J.Mater. Sci.*, **19**, 1767-1976 (1984).

EFFECTS OF DISSOLUTION AND EXSOLUTION OF Ni IN YSZ

Soren Linderoth and Nikos Bonanos
Materials Research Department
Risoe National Laboratory
DK-4000 Roskilde
Denmark

ABSTRACT

The effect of the dissolution of Ni in zirconia, doped with about 8 mol% yttria, has been investigated with respect to sinterability, grain growth, solubility and electric conductivity. The solubility is in the 1 at% range and the addition of such small amounts of NiO causes the zirconia to sinter at temperatures about 65°C lower than without NiO present. The grains become about twice as large. At higher dopant levels the NiO inhibits grain growth. The conductivity at low temperatures is lowered when adding NiO, while at 850°C the effect is found to be positive up to the solubility limit. Upon reduction the conductivity drops dramatically.

INTRODUCTION

A porous Ni/yttria-stabilized zirconia (Ni/YSZ) cermet is the most commonly used anode in solid oxide fuel cells [1]. The anode may be fabricated by spraying a slurry of a NiO/YSZ mixture onto a sintered YSZ electrolyte, followed by an ad-sintering at elevated temperatures, or vice versa, the electrolyte may be applied onto an NiO/YSZ support and sintered. During the ad-sintering, which is typically performed at temperatures of about 1100-1500°C, Ni^{2+} may dissolve in the YSZ. Dissolution of Ni^{2+} will give rise to lattice parameter changes and thereby cause mechanical stresses with subsequent possibilities for failure. The dissolution may also affect the phase stability [2] and the conductivity of YSZ [3]. The solubility of NiO in 6-10 mol% YSZ doped is less than 2 mol% [4, 5]. The solubility, at 1600°C, varies only a little within the investigated range of yttria concentrations [5]. In this work, the impact of dissolved NiO in cubic 8YSZ on the sinterability, grain size and conductivity, both in air and a reducing atmosphere, have been investigated.

EXPERIMENTAL

A commercially available TZ8Y (Tosoh, Tokyo, Japan) powder and an ultrafine NiO powder prepared by plasma synthesis [6] were used for the experiments. The powder mixtures were ball milled for 24 hours in ethanol using zirconia balls and a plastic container. Powder slurries were dried at 80°C and then crushed and screened through a 160 µm sieve. The resultant powders were uniaxially pressed into pellets with a diameter of 12 mm and height 10-12 mm with a pressure of 100 MPa and sintered in air in the temperature interval 1200-1500°C for 2 and/or 100 hours and at 1600°C for 12 hours. Temperature increasing and decreasing rates were 2 and 5 Kmin^{-1}, respectively. Selected samples were also sintered for 80 h at 1600°C and quenched into water. During sintering the pellets were placed on top of well-sintered pieces of 8YSZ. For dilatometric measurements uniaxially pressed pellets were additionally cold isostatically pressed with a pressure of 200 MPa. NiO-free samples were prepared in a similar manner as the NiO-doped samples. Table I shows the composition and the cubic lattice parameter of the investigated samples after heat treatment in air for various temperatures and times.

The x-ray diffraction (XRD) investigations were performed by using Cu-K$_\alpha$ radiation in transmission mode in a STOE STADIP diffractometer equipped with a Ge monochromator and a position sensitive detector. A standard Si powder was used as an internal standard for the lattice parameter calculations. NiO content in YSZ-NiO solid solutions was estimated from the lattice parameter changes of the cubic zirconia [7]. The lattice parameter of YSZ-NiO solid solutions was estimated from the profile fitting results of the selected interval of $2\theta = 68-80°$ containing two Si maxima, (400) and (331), and one YSZ maximum, (400). Profiles were fitted with a pseudo-Voigt function. A comparison showed that the lattice parameters determined from a Rietveld analysis and the profile fitting differ less than 0.00006 nm.

Table I. NiO content for different samples and lattice parameter (nm) of cubic zirconia for various for various heat treatments. The yttria content of the zirconia was 7.83 mol%.

Sample	NiO (at%)	1200°C 2h	1500°C 2h	1200°C/100h after 1500°C/2h	1600°C 12h	1600°C 80h, quench
T0	0	0.51422	0.51401	0.51411	-	-
T75	0.75	0.51395	0.51378	0.51382	0.51374	0.51378
T15	1.5	0.51391	0.51358	0.51367	0.51349	0.51350
T25	2.5	0.51393	0.51359	0.51369	0.51340	0.51345
T10	10	-	-	-	-	-

The microstructure of sintered samples after polishing was examined by scanning electron microscopy (JSM-5310LV). The grain size was estimated from the average grain area assuming spherical grains.

Impedance spectroscopy was used to resolve the grain interior and grain boundary components of resistance. Impedance spectra were obtained on discs of 8 mm diameter with platinum paste electrodes, in the temperature range 300 to 850°C, in flowing dry air. A *Solartron 1260* impedance analyser was used in two-terminal four-wire arrangement, in order to minimise the effects of lead resistance and inductance. The impedance data were analysed using the software *Equivcrt* [8].

RESULTS AND DISCUSSION

The phase diagram of the ZrO_2-Y_2O_3 system is controversial with regard to the cubic (c) phase boundary at temperatures 1000-1500°C [9]. The appearance of the t-phase in the XRD patterns after the sintering at 1200°C for 100 h, as well as increase of the cubic lattice parameter, shows the given composition is in the (c+t)-phase field at 1200°C. The presence of a low-yttria containing t-phase will result in a higher-than-average yttria concentration in the c- phase at 1200°C. Therefore, the lattice parameter will be higher [10] than that observed after sintering at 1500°C. This transformation process of the cubic phase of 8YSZ during annealing is one reason for the degradation of the ionic conductivity during a long-term testing at 1000°C [10]. The long-term heat treatment at 1200°C can be considered as an accelerated aging test.

The NiO-doped samples showed slightly different phase composition than the NiO-free ones. Only traces of the t-phase could be detected after sintering at 1200°C for 2 h. This may be ascribed to the fact that NiO together with yttria acts as a stabilizer of the c-phase [2]. The t-phase is not detected by XRD for sintering temperatures above 1300-1350°C or after extended annealing at 1200°C for the NiO-doped samples.

Dilatometric curves of five investigated samples are shown in Figure 1. It is seen that doping with as little as 0.75 mol% NiO makes the onset of sintering decrease by about 65°C. The sintering of the NiO-doped samples starts around 1000°C while the dissolution of NiO in the YSZ was found to begin at about 1200°C [5]. This means that the lowering of the onset temperature of sintering is most likely caused by an enhanced surface diffusion. Some difference in shrinkage of the NiO-doped and non-doped samples seen in Fig. 1 may be caused by a different relative density of the prepared samples.

In Figure 2 is shown the grain size and the dissolved NiO for the samples sintered at 1500°C as a function of the overall NiO content. The averaged lattice parameters for slowly cooled and quenched samples from 1600°C were used to

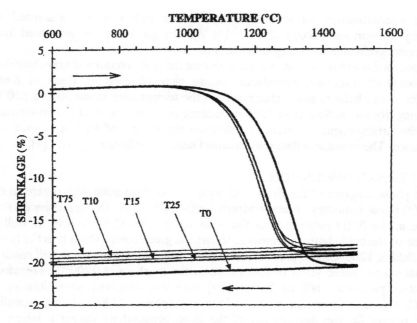

*Figure 1. Dilatometric curves for various overall NiO doping
(see Table I).*

calculate the dissolved amount of NiO in the YSZ lattice by the use of the relationship [5]:

$$a_{\text{YSZ-NiO}} = 0.5119 + 0.0001394\, m_{\text{YO}_{1.5}} - 0.000515\, m_{\text{NiO}} \qquad (1)$$

where $a_{\text{YSZ-NiO}}$ is the lattice parameter of the YSZ-NiO solid solution and $m_{\text{YO}_{1.5}}$ and m_{NiO} are the concentrations of $\text{YO}_{1.5}$ and NiO in mol%. At a total NiO concentration of 10 mol% the dissolved NiO amount in the YSZ lattice is about 1.1 mol%. It is noteworthy that the dissolved NiO amount is less than that observed in the case of YSZ-NiO composites prepared by the plasma synthesis method (Fig. 3). The higher solubility of NiO in that case is most likely due to the smaller YSZ particle size of the plasma prepared powders, and hence a higher interphase contact area.

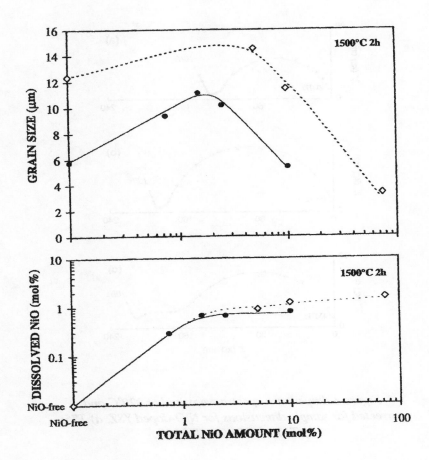

Figure 2. Grain size and dissolved amount of NiO as a function of overall NiO content. Solid dots show results from the present work while diamonds are taken from Ref. 5.

The grain size of the YSZ is seen (cf. Figure 2) to increase for NiO contents below the solubility limit, reaching a maximum at around the solubility limit, whereafter the grain size decreases with increasing the overall NiO content. A very similar behaviour has been observed previously for ultrafine YSZ-NiO composites [5] (see Figure 2). It is clear that above the solubility limit undissolved NiO

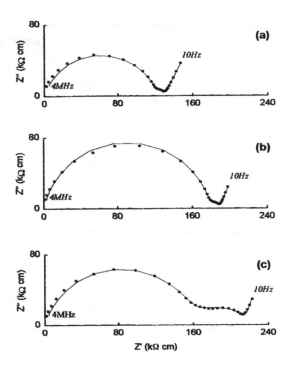

*Figure 3. Impedance spectra measured at 320°C and
corrected for sample dimensions for NiO-doped YSZ. a) T75;
b) T15; c) T10 (see Table I).*

particles in YSZ matrix act as inhibitors of the grain growth. This is due to a requirement for additional energy to move the boundary from the inclusion [11]. The grain growth enhancement below the NiO solubility limit might be a result of an enhanced lattice diffusion since replacement of Zr^{4+} by Ni^{2+} causes the formation of two oxygen vacancies in the YSZ lattice.

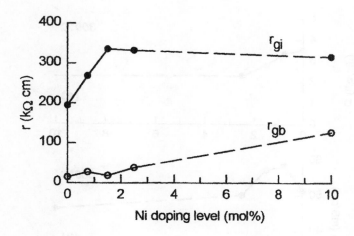

Figure 4. Dependence of the grain (r_{gi}) and the grain boundary (g_{gb}) resistivity components at 300°C on doping level for NiO-doped YSZ.

Figure 3 shows impedance plots for three NiO-YSZ samples obtained at 320°C. From these plots the contribution to the resistance from the grain interior and the grain boundaries can be resolved. At elevated temperatures it is more difficult to distinguish these properties. The deduced grain and grain boundary resistances, at 300°C, are reproduced in Figure 4 as a function of overall NiO content in the samples. The larger contribution comes from the grain interior and is seen to increase by about 75% from the undoped sample to about the solubility limit. Above the solubility limit no change is seen for the interior, as should be expected. The grain boundary resistance increases with increasing NiO content. In Figure 5 is shown the total conductivity as a function of overall NiO content at 300°C and 850°C. As can also be deduced from Figure 4, the total conductivity decreases upon NiO doping at 300°C, while at 850°C the conductivity reaches a maximum at about the solubility limit. The increase for low NiO content may be ascribed to the the phase stabilisation of the cubic phase by the NiO, as discussed previously. At higher NiO contents the negative effect of NiO on conductivity in the grains becomes most important.

The conductivity in an atmosphere where the NiO will be reduced to Ni has been found [3] to lower the total conductivity substantially. The conductivity may

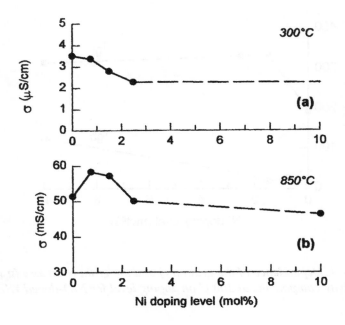

Figure 5. Dependence of the total conductivity of NiO-doped YSZ at two temperatures.

decrease by as much as 50% at 1000°C. Figure 6 shows impedance spectra of a reduced sample (1.5 at% Ni).

CONCLUSIONS

The solubility of NiO in cubic zirconia doped with 8 mol% Y_2O_3 (8YSZ) is found to be about 1 mol% at 1600°C, resulting in a decrease of the lattice parameter by approximately 0.06%. The onset temperature of sintering decreases by about 65°C by the addition of 0.75 at% of NiO to 8YSZ. Higher NiO contents has no further effect on the sintering profile. The average grain size after sintering at 1500°C reaches a maximum around the solubility limit where the grain size is about twice that of the NiO-free sample. The presence of NiO in the YSZ causes

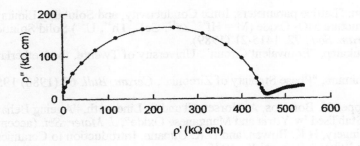

Fig. 6. Impedance spectrum measured at 320°C after reduction of NiO to Ni. The sample is T15.

the total conductivity to decrease at low test temperatures, while low NiO contents may be beneficial at higher temperatures, probably due to a stabilisation effect. In a reducing atmosphere the conductivity of the YSZ decreases substantially due to a destabilisation of the cubic phase.

ACKNOWLEDGMENTS

The work has been in part been supported through the Joule/Thermie project IDUSOFC (EU contract JOE3-CT-0005) and the Danish SOFC Program by the Danish Energy Agency.

REFERENCES

[1] N.Q. Minh, "Ceramic Fuel Cells" *J. Amer. Ceram. Soc.* 76, 563-88 (1993).

[2] A. Kuzjukevics and S. Linderoth, "Influence of NiO on Phase Stabilization of 6 mol% Yttria-Stabilized Zirconia", *Mater. Sci. Eng.,* A232, 163-167 (1997).

[3] S. Linderoth, N. Bonanos, K.V. Jensen and J.B. Bilde-Soerensen, "Effect of NiO-to-Ni Transformation on Conductivity and Structure of Yttria-stabilised ZrO_2", *J. Amer. Ceram. Soc.* (accepted).

[4] A. Kuzjukevics, S. Linderoth, and J. Grabis, "Plasma Produced Ultrafine YSZ-NiO Powders" *in:* Proc. of the 17th Riso Intern. Symp. on Materials Science "High Temperature Electrochemistry: Ceramics and Metals", edited by F.W. Poulsen, N. Bonanos, S. Linderoth, M. Mogensen, and B. Zachau-Christiansen (Riso National Laboratory, Roskilde, Denmark, 1996), p.319-324.

[5] A. Kuzjukevics and S. Linderoth, "Interaction of NiO with Yttria-Stabilized Zirconia", *Solid State Ionics,* 93, 255-261 (1997).

[6] J. Grabis, A. Kuzjukevics, D. Rasmane, M. Mogensen, and S. Linderoth, "Preparation of Nanocrystalline YSZ Powders by the Plasma Technique", *J. Mater. Sci.* 33, 723 (1998).

[7] D.-J. Kim, "Lattice parameters, Ionic Conductivity, and Solubility Limits in Fluorite- Structure MO_2 Oxide ($M = Hf^{4+}, Zr^{4+}, Ce^{4+}, Th^{4+}, U^{4+}$) Solid Solutions" *J. Amer. Ceram. Soc.*, 72, 1415-21 (1989).

[8] B.A. Boukamp, "Equivalent Circuit", University of Twente, the Netherlands (1989).

[9] M. Yoshimura, "Phase Stability of Zirconia", *Ceram. Bull.* 67 (1988) 1950-1955.

[10] C.C. Appel, N. Bonanos, A. Horsewell and S. Linderoth, "Ageing Behaviour of Zirconia Stabilised by Yttria and Manganese Oxide", *J. Mater. Sci.* (accepted).

[11] W.D. Kingery, H.K. Bowen, and D.R. Ulmann. Introduction to Ceramics, 2nd edition, John Wiley & Sons, N.Y., 1976.

MULTILAYERD CERAMIC REACTOR FOR THE STEAM REFORMING OF METHANOL INTO HYDROGEN ENRICHED GAS

Don Gervasio, Stephen Rogers, Ramesh Koripella, Sonja Tasic, Daniel Zindel, Rajnish Changrani, Christopher K. Dyer*, Jerry Hallmark and David Wilcox
Motorola Labs
7700 South River Parkway, Tempe, AZ 85284
*Current address, Stevens Institute of Technology, Hoboken, NJ

ABSTRACT

Monolithic ceramic structures with a high degree of embedded system controls and integrated functionality can be fabricated utilizing low pressure ceramic lamination technology. This paper describes the design and development efforts of a miniature methanol steam reformer for man-portable fuel cell applications. A small (35 x 15 x 5mm) microreactor with integrated fuel vaporizer and embedded thick-film heaters was designed and fabricated using this technology. The effects of reactor temperature, liquid fuel flow rate, and methanol/water ratio were studied to determine the influences on methanol conversion and reformate output gases H_2, CO, and CO_2 for a particular reformer design.

INTRODUCTION

A miniature hydrogen generator is being developed for an elevated temperature (150 to 225°C) fuel cell [1,2] which is capable of efficiently converting the reformer's hydrogen rich gaseous product into electrical energy. The electrical energy results when hydrogen is supplied to the anode and air is supplied to the cathode of the fuel cell. This reformer/fuel cell system is designed to power portable electronics much longer than currently available batteries.

The hydrogen generator is a chemical reactor, called a steam reformer, which converts liquid methanol and water to hydrogen rich gas at elevated temperatures (200 to 300°C), according to the reversible endothermic chemical reaction 1 given below.

$$CH_3OH + H_2O \rightleftharpoons CO_2 + 3H_2 \qquad \text{Reaction 1}$$

Ceramic structures are well suited for housing a steam reformer. Steam reforming is a catalyzed reaction employing heterogeneous metal and metal oxide catalysts dispersed on a ceramic support [3,4,5]. Ceramics are used as the catalyst support, because ceramics do not corrode or interfere with the catalysts during the harsh reforming conditions, namely, flowing water and methanol at high temperatures. Some ceramic supports have even been found to activate or favor the desired reforming catalyses.

Structures formed from co-fired multilayered ceramic substrates (often called CMEMS in analogy to conventional silicon-based microelectromechanical structures, MEMS) permit the introduction of diverse structural features into a ceramic monolith, and CMEMS also permits ready integration of subassemblies, ceramic or otherwise. The diversity of structural features include: vaporization and reaction chambers, gas and liquid flow channels, air-gaps for making thermal and electrical insulators, and metal filled vias for making thermal and electrical conductors. Possible subassemblies include: pumps, valves, thermistors, thermocouples, fuel reservoirs and electronic controller boards. Ceramics are well suited for housing most if not all of the various components of the high temperature portable power supply including: the reformer, the fuel cell, the control electronics, fuel storage, pumps and fans, etc. The present discussion focuses on exploiting the CMEMS approach to make a methanol to hydrogen steam reformer.

EXPERIMENTAL

CMEMS construction of steam reformer: A prototype miniature steam reforming device was designed and fabricated using the multilayer ceramic technology. A commercial LTCC (low temperature co-fired ceramic) tape material (DuPont 951AT - Green Tape) was used in the fabrication of the device. Multilayer ceramic technology is widely used in designing various electronic ceramic components. It is also being used for ceramic MEMS applications because of its ability to form 3D structures with microfluidic channels and interconnects, and its ability to integrate various functionalities in the device. The green tape consists of ceramic (alumina, glass and other inorganic additives) particles within a polymer matrix and set on a Mylar backing. In addition to the ceramic particles, the green tape contains binders, solvents, dispersant, plasticizer, oxides and glass. Prior to the firing/sintering of this material, it is very flexible and can be punched to form holes and channels. Microchannels and the catalyst-containing reforming chamber were formed this way on individual layers and stacked together and laminated to form a monolithic 3D structure. Screen-printing technology was used to form thick film resistors, which were used as embedded electrical heaters to provide the required reforming temperatures of 200-300°C. The green laminate was then fired at 850°C to form a hard monolithic ceramic device.

Reformer internal structure and catalysts: Two chambers were created inside the device, one acting as a fuel vaporizer and the other as a reforming chamber. The reformer cavity was filled with a commercial $Cu/ZnO/Al_2O_3$ powder and sealed gas tight using a high temperature alumina adhesive. Glass tubes were connected to the inlet and outlet portions of the ceramic device using the same adhesive cement. Precise flows of liquid methanol/water mixed at various mole ratios were pumped using a syringe pump through glass tubes (id=0.53mm, od=0.8mm, l=55mm) into the vaporizer chamber, which was maintained at 145 to 180°C by an embedded thick film electrical heater. Fuel vapors were then transported into the reforming chamber using internal channeling. The steam-reforming chamber was also heated by the embedded thick film electrical heater. Output from the reformer chamber was transported through a glass tube, which was connected to a GC for analyzing the output gases. In the current set of experiments reported in this paper, approximately 50 mg of catalyst power was used. The reforming temperature was varied from 180-230°C. The vaporizer/reformer was fed a liquid with a methanol/water mixture spanning 0.95:1 and 1:1.05 mole ratios and at a flow rates between 1 to over 10 microliter/min.

Gas Chromatographic Analysis of Reactor Effluent: The gas effluent from the reforming reactor was analyzed using capillary gas chromatography on a Hewlett-Packard 6890 gas chromatograph (Agilent Technologies, Palo Alto, Calif.) equipped with a thermal conductivity detector (TCD) and a 30m x 0.53mm id Carboxen-1006 PLOT column (Supelco, Bellefonte, PA) with helium carrier gas set for constant flow condition of 4 ml/min at 40°C. A heated two-position six-port valve contained in an insulated housing allowed the ceramic reactor to be coupled to the analytical system or out to a digital flowmeter or fuel cell with minimal disruption to the steam reforming occurring in the reactor. Reformate gases (hydrogen, carbon monoxide, carbon dioxide, water, and methanol) were separated using a GC algorithm, namely, 2 minutes effluent flow through a 10ul gas sample loop at 125°C; gas sample valve actuation for injection of the gas sample onto the PLOT column at 40°C; GC oven programmed for 2 min at 40°C, ramped at 50°C/min to 200°C, held at 200°C for 5 minutes. Prior to analyzing the reactor output gases, calibration runs were made on the GC using standard gas mixtures of CO, CO_2, and H_2.

DISCUSSION AND RESULTS

Steam Reforming in the miniature ceramic micro reactor: The ceramic reactor used in these experiments is shown below in Fig 1. This figure is a cut away section of the reactor with the top removed showing the fuel inlet, fuel vaporizing chamber, steam reforming chamber filled with the catalyst, internal interconnect between these two chambers and the gas outlet.

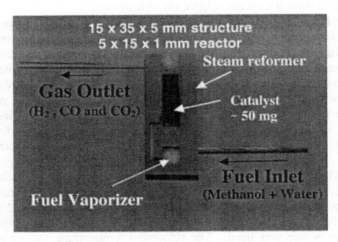

Figure 1. Cut away (lid removed) of the ceramic microreactor for the steam reforming of methanol to hydrogen enriched gas.

The dimensions of the vaporizer were ~ 5mm x 5mm x 1mm thick, and the steam reformer were ~ 5 x 15 x 1 mm thick. The vaporizer chamber was connected to the reformer chamber via a single channel to the reactor chamber. The reformer cavity contained a packed bed of ~ 50 mg of $Cu/ZnO/Al_2O_3$ of particle catalyst (~20 micron particle size). All of these features were housed within a ceramic structure, which had external dimensions of 15 x 35 x 5 mm.

Fig. 2 shows a typical GC detector response for output from the reformer operating at 180, 200 and 230°C with a liquid feed to the vaporizer containing 1 mole MeOH and 1.05 mole H_2O fed at 5 microliters/minute. This graph shows three different GC runs for the three temperatures overlaid. Area under the peaks is used in quantification of the gas species. As seen from these results, all the gas species expected from the methanol steam reforming reaction can be separated and detected using our GC set up. In order to determine the relative concentration of the same species (the peak area at a given retention time) as a function of temperature, the direct comparison of the TCD response could be used. On the other hand, to determine the relative concentrations of different species (i.e., to make a comparison of a peak area at one retention time to the peak area at another retention time), the peak area at a given retention time had to be calibrated to a known concentration of gas which gave a peak at that given retention time. This calibration was done and the results are shown in Figure 3 discussed next.

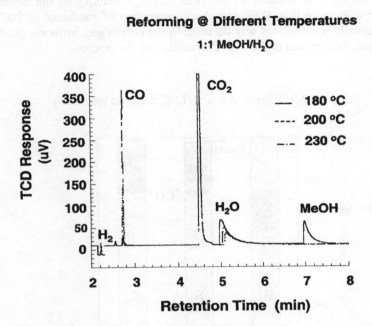

Reforming @ Different Temperatures

1:1 MeOH/H₂O

Figure 2. Response of thermal conductivity detector (TCD/microvolts) in gas chromatograph while separating an aliquot of gas output from ceramic microreformer when operated at 180, 200 and 230°C. Liquid Feed to vaporizer contained 1 mole MeOH and 1.05 mole H2O and was fed at 5 microliter/min. Area under peak at a given retention is proportional to the concentration. Calibration standards were used to compare peaks at different retention times for relative concentrations.

The performance of the reformer was investigated over a range of conditions including input feeds with mole ratios of $H_2O/MeOH$ from 0.95 to 1.05 and input flow rates from 1 to over 10 microliter/minute. With the present reactor design, a liquid feed with a mole ratio of 1.05 $H_2O/MeOH$ flowed into the reactor at up to 10 microliters per minute proved well suited for the making reformate gas that was suitable for supplying the elevated temperature fuel cell.

Fig. 3 summarizes the effects of temperature on reactor performance. Fig. 3 shows the compositions of the gases which came out of the reactor, when it was operated from 180 to 230 °C while being continuously fed 5 microliter/minute of a liquid fuel consisting of 1 mole of methanol and 1.05 mole of water. As seen from these results it is feasible to perform steam reforming in the ceramic microreactor with more than 90% extent of conversion of methanol at 200°C. At 230°C, virtually no methanol was detected in the output gas, however the CO content increased compared to the 200°C operation of the reactor.

(mol MeOH/mol Water :1/1.05, 5 ul/min inlet fuel)

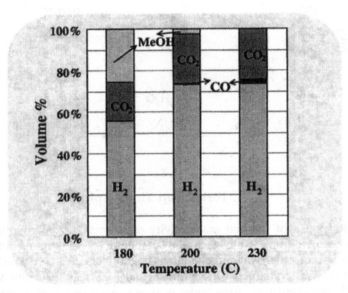

>90% MeOH Conversion @ 200C

Figure 3. Vol% of H_2, CO_2, CO and MeOH from reactor at 180, 200 and 230 °C. Liquid feed Rate = 5 micro-liter/minute; feed composition, 1 mole of MeOH and 1.05 mole of water, 50 mg catalyst. Gas vented to ambient pressure (exhaust id=0.53mm, l=55mm).

The main product distribution is consistent with >90% extent of conversion of the methanol to carbon dioxide and hydrogen as given in reversible endothermic reaction 1, above. The small amounts of the undesired CO can come from the partial decomposition of methanol, as shown in reaction 2, below.

$$CH_3OH \longrightarrow CO + 2H_2 \qquad \qquad \textbf{Reaction 2}$$

This is not likely at lower reforming temperatures (200 to 300°C), but can be prominent at higher temperatures (>400°C). At lower reforming temperatures, the CO likely results from a reversible reaction, the so-called "reverse water gas shift" reaction, as shown in reaction 3, below.

$$CO_2 + H_2 \rightleftharpoons CO + H_2O \qquad \qquad \textbf{Reaction 3}$$

The products of steam reforming reaction, CO_2 and H_2, become the reactants for the reverse water gas shift reaction. As the temperature increases and pressure decreases, reaction 3 is less favored in the forward direction, and vice versa. Nevertheless, the rate of all reactions becomes higher at higher temperatures, and so some CO is expected to result as temperatures go much above 200°C. This is consistent with the experimental results summarized in Fig. 3 above.

The main objective of steam reforming of methanol (to H_2 enriched gas) is to retain a significant fraction of the relatively high energy density of methanol (4700 Wh/liter for pure MeOH; 3500 Wh/liter for 1MeOH:1H_2O mixture) while converting it to a fuel (H_2 gas) which can supply a fuel cell in order to generate electrical power to extend the lifetime of portable electronics over that obtained when using the best batteries (e.g., Li-ion provides 200 Wh/liter). A 25% chemical to electrical conversion efficiency would offer a marked improvement over presently available batteries.

Fig. 4 shows the results from the continuous operation of the reactor for 120 hours at 230°C with the 1:1.05 mole ratio of MeOH/H_2O fuel at 10 microliter/min flow rate into the reactor. The concentrations of the CO and (CO plus MeOH) are both plotted in this figure as a function of reactor operating time. At the times shown in the figure (38hr, 59hr and 106hr) the reactor was backflushed with a clean He gas while the reactor was maintained at the same temperature. After 68hrs the fuel solution in the syringe pump ran out and a fresh solution was inserted in the syringe pump without disrupting the operation of the reactor. As seen from the data, after 80hr of testing the methanol content in the output gases increased exponentially while the low CO content was maintained. This indicates the degradation of the catalyst performance for steam reforming reaction in the reactor. Cu/ZnO/Al$_2$O$_3$ catalyst is a well-studied catalyst for the steam reforming reaction with good stability and longevity. The degradation of performance observed in our results is more likely related to the miniature reactor design and operating conditions.

As long as the combined volumes of CO and MeOH is less than 3 to 5% of the total volume of the hydrogen enriched gas output by the reforming reactor, the fuel cell is operable. Exceeding 3 to 5% signals the end of the useful lifetime of the reformer.

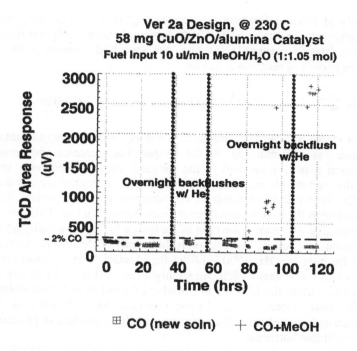

Figure 4.   Response in time of TCD signals for CO and MeOH coming out of reformer reactor (Version 2a). Note: 250microV signal corresponds to ~ 2% total CO. The signal in microV is linearly related to %CO in gas. Useful lifetime of this reactor is ~ 95 hours.

Looking at Fig. 4 above it can be seen that our current design of the miniature ceramic reforming reactor produced a usable output gas up to about 95 hours.

There are a number of factors, which can cause the reformer to fail. One reason for failure is catalyst deactivation either from air oxidation (reversible?)

or dealloying (irreversible) due to inadvertent soaking by fuel leaving residual corrosive moisture. Mechanical deactivation is also a possibility.

Catalyst particles can clog the corners of the singular channel connecting the fuel vaporizer to the reformer chamber (see Fig. 1). Particles could have been carried into this channel by liquid fuel which was inadvertently admitted into the interior of the reformer during the reformer shut down and start up procedure, used as the reformer was discontinuously tested over several days. The pressure in the working device is expected to only be a fraction of a psi to no more than 3 psi. This conclusion is based on the observed pressure drop determined manometrically as dry N_2 gas was flowed through the reactor at various N_2 gas flow rates, and extrapolating to the typical reformate gas flow rate of 10 sccm for our application. Accordingly, the effect of the average reactor pressure on the performance of the reforming reactor is expected to be negligible.

However, sometimes pulsing of the inlet fuel was observed during operation. Such pressure pulses could possibly cause channels to form in the catalyst bed. X-ray images of reactors were obtained before and after such pulsing occurred. Channels were not observed in unused reactors but were observed after pulsing occurred in a reactor. The effects of channels in the bed are not expected to cause a major change in performance. However since channeling has been observed, the effects of channels on reactor performance is one subject to be examined as part of planned reformer modeling work. Most of the problems in the present reforming reactor design seem to come about either because there is only one channel between the vaporizer and reformer or because liquids entered the reactor. In order to maximize reformer lifetime, a new reactor design has been developed with a multichannel fluid conduit between the vaporizer and the steam-reforming chamber. Work is now focused on this new design taking care to keep the interior of the reforming chamber free of liquids to test if the lifetime of the reactor can be extended.

CONCLUSION

An example of a monolithic ceramic structure has been presented for converting methanol to hydrogen-enriched gas. Fabrication of the structure utilized multilayer ceramic (LTCC) technology. This ceramic technology readily allowed construction of a device, which contained embedded thick-film heaters, fluid flow channels, a fuel vaporizer and a miniature catalytic reaction chamber all within ~35 x 15 x 5 mm outside dimensions for steam reforming methanol into hydrogen enriched gas. The steam reformer performance was characterized in order to determine the effects of reactor temperature, liquid fuel flow rate, composition of the fuel (water/MeOH ratio) on the extent of conversion of the fuel and the distribution of H_2, CO and CO_2 in the product gas. The miniature ceramic reactor converted greater than 95% of the fuel to hydrogen enriched gaseous product, and this reformer product gas was found suitable (< 3% CO and/or methanol vapor) for operating an elevated temperature (150 to 225°C) fuel cell. Future work will focus on using the CMEMS approach as a convenient means to rapidly modify reformer design and exploring the results of the planned modifications.

REFERENCES

[1] Gervasio, D., Razaq, M., Razaq, A., Yeager, E.B., "Nafion 117 Membrane with Concentrated Phosphoric Acid as the Proton Solvating Agent for Use in a Solid Polymer Electrolyte (SPE) Fuel Cell", Electrochem. Soc. Spring Meeting, Washington, D.C., Ex. Abs., 91-1, p. 13, 1991.

[2] Wang, J-T., Savinell, R. F., Wainright, J. S., Litt, M., and Yu, H., "A H2/O2 Fuel Cell using Acid Doped Polybenzimidazole as a Polymer Electrolyte", Electrochimica Acta, Vol. 41 , pp. 193-197, 1996; Weng, D., Wainright, J. S., Landau, U., and Savinell, R. F., "Electro-osmotic Drag Coefficient of Water in Polymer Electrolytes at Elevated Temperatures", Journal of the Electrochemical Society, Vol. 143, pp. 1260-1263, 1996.

[3] Amphlett, J., Creber, K., Davis, J., Mann, R., Peppley, B., Stokes, D., "Hydrogen Production by steam Reforming of Methanol fo Polymer Electrolyte Fuel Cells", Int. J. Hydrogen Energy, Vol. 19, No.2, pp. 131-137, 1994.

[4] Peppley, B. Amphlett, J., Kearns, L., Mann, R., "Methanol-steam reforming on Cu/Zn/Al2O3. Part1: the reaction network", Applied Catalysis A: General 179, pp. 21-29, 1999.

[5] Charles N. Satterfield, Heterogeneous Catalysis in Practice, McGraw Hill, NY, NY (1980). See p. 294 on low temperature shift catalysis and Ch. 10 for a general discussion of catalytic steam reforming processes and reactors.

SiO_2-P_2O_5-ZrO_2 SOL-GEL / NAFION^{TM} COMPOSITE MEMBRANES FOR PEMFC

M. Aparicio and L.C. Klein
Department of Ceramic & Materials Engg
Rutgers, The State University of New Jersey
607 Taylor Road, Piscataway, NJ 08854-8065

K.T. Adjemian and A.B. Bocarsly
Department of Chemistry
Frick Laboratory
Princeton University,
Princeton, NJ 08544

ABSTRACT

One of the major limitations of current proton-exchange membrane fuel cells (PEMFC) is that the Pt anode electrocatalyst is poisoned by CO. While it is desirable to operate a PEMFC at temperatures above the standard boiling point of water, both to reduce the tendency for CO poisoning and for more efficient water and thermal management, membranes lose conductivity because of drying. The incorporation of SiO_2-P_2O_5-ZrO_2 gels into NAFION^{TM} can improve its proton conductivity and thermal stability above 100°C. Composite membranes were prepared by an impregnation of NAFION^{TM} with SiO_2-P_2O_5-ZrO_2 sol. Composite membranes had an average weight increase of about 10% after drying at 80-150°C. The membranes were characterized by TGA/DTA, SEM-EDX and functional testing in a fuel cell.

INTRODUCTION

Fuel cells are being developed for use in transportation, owing to their inherently high-energy efficiency and ultra low or zero emissions of environmental pollutants, compared to internal combustion engines. Recent advances have made proton-exchange membrane fuel cells (PEMFC) a leading alternative to internal combustion and diesel engines for transportation. These advances include the reduction of the platinum electrode catalyst needed, and membranes with high specific conductivity, good water retention and long lifetimes (1). A major limitation of the current PEMFC is that the Pt anode electrocatalysts is poisoned by CO at the 5 to 10 ppm level in the state-of-the-art fuel cells operating at about 80°C. CO-tolerant electrocatalysts (e.g., Pt-Mo, Pt-Ru) have been investigated to enhance CO tolerance, but the problems with these electrocatalysts are that (i) the Pt loading is 5 to 10 times higher than required for pure platinum catalyst and (ii) their CO tolerance level is about 50 ppm and even at this level this is an increased overpotential for the anodic reaction (2,3). Research has also been conducted on oxidizing the CO adsorbed on the Pt electrocatalyst by injecting a small amount of air (<2%) into the fuel stream entering the anode chamber or by adding minimal amounts of H_2O_2 to the humidification bottles of the reactant gases (4,5). Another approach is to take into consideration that the absolute free energy of adsorption of CO on Pt has a larger positive temperature dependence than that of H_2 which means that the CO tolerance level should increase with temperature (6).

The current PEMFCs also are complicated by the water management. For example, the proton conductivity of the PEMFC increases linearly with the water content of the membrane, with the highest conductivity corresponding to a fully hydrated membrane. It is for this reason that the reactant gases are appropriately humidified before entering the cell. While it is desirable to operate a fuel cell at a temperature above the boiling point of water from the standpoint of increased reaction kinetics and lower susceptibility to CO poisoning, the membranes lose conductivity due to drying. Membrane dehydration also causes the membrane to shrink which reduces the contact between the electrode and membrane, and may also cause pinholes to form leading to the cross-over of the reactant gases. The problem is that the vapor pressure of water increases very rapidly with temperatures above 100°C. In order to maintain the needed hydration for the polymeric membrane and the partial pressures of the reactant gases to attain the desired performance levels, the total pressure has to be increased significantly.

To solve both the CO poisoning and the water-thermal management problems, the present state-of-the-art proton exchange membranes need to be modified in order to remain hydrated at higher operating temperatures for the PEMFCs. There have been previous efforts to enhance the water retention of Nafion and related membranes by incorporation of hydrophilic metal oxides (e.g. SiO_2). Difficulties have been encountered because the metal oxide particles are micron size and are not sufficiently small to enter the nanopore structure of the membrane (7). To overcome this problem other authors used a sol-gel technique to introduce a polymeric form of oxides into the perfluorosulfonic acid. Using this method, it was shown that the oxides enter the fine channels (~5 nm in diameter) (8).

We propose to enhance the proton conductivity of ion-exchange membranes at elevated temperatures by incorporating the membrane with sol-gel processed silicophosphates. These gels are designed to provide a high concentration of protons by tying up the water in the pores and reducing its volatility. Silicophosphate gels have been shown to be fast proton-conducting solids. The mobility of protons increases when the protons are strongly hydrogen-bonded. Compared with Si-OH, phosphate glasses are more efficient for high protonic conduction because the hydrogen ions are more strongly bound to the non-bridging oxygen. Also, the hydrogen in the P-OH group is more strongly hydrogen-bonded with water molecules, resulting in an increase in the temperature necessary to remove the water from P-OH. At the same time, the presence of the silicate network is important because of its mechanical strength and chemical durability. The introduction of cations such as Zr^{4+} into silicophosphate gels results in improved chemical stability (9-11). The sol-gel process is suitable in this case because the low surface tension and viscosity of the sols allows infiltration into Nafion membranes, along with the preparation of microporous gels to improve the water retention through hydrogen-bonded protons on hydroxyl groups.

EXPERIMENTAL

$60SiO_2$-$30P_2O_5$-$10ZrO_2$ sol was prepared by the sol-gel method using tetraethyl orthosilicate [$Si(OC_2H_5)_4$, TEOS] from Fisher Scientific, triethyl phosphate [$PO(OC_2H_5)_3$, TEP] from Aldrich Chemical, and zirconium-n-propoxide [$Zr(OC_3H_7)_4$, TPZr] from Johnson Matthey Electronics as starting materials.

The sol was prepared by mixing two solutions. Solution A was prepared by mixing TEOS, half the volume of propanol (solvent) from Fisher Scientific, TEP, HCl from Fisher Scientific (to control the pH at ~2) and water [molar ratio of water/(TEOS+TEP)=2] at room temperature and stirred for 1 h. Solution B was prepared by mixing TPZr, the other half of the propanol and acetylacetone from Fisher Scientific (molar ratio of acetylacetone/TPZr=1) at room temperature and stirred for 1 h. Both solutions were subsequently mixed together and stirred for another 1 h. The remaining amount of water was added drop by drop, and then the solution was stirred for 15 min. The sol has a concentration of 70 grams of solid per liter, and a final molar ratio of water/precursors=5.5.

The preparation of the infiltrated Nafion consisted of first pre-treating the Nafion 115 membrane with 3% by vol. H_2O_2 for 2 hours at 80°C, followed by 50% by vol. H_2SO_4 for 2 hours at 80°C. The membrane is then treated three times in distilled H_2O at 80°C to remove any excess acid. After drying the membrane for 3 days at 80°C, it was immersed in the $60SiO_2$-$30P_2O_5$-$10ZrO_2$ sol for 3 hours. Then, the membrane surfaces were cleaned with propanol to avoid the formation of surface-attached silicate layers. After the treatment, the membrane was placed at room temperature for 5 hours and then in an oven at 80-150°C for 2 days. The dried infiltrated samples had an average weight increase of about 10%.

The thermal behavior of original Nafion and Nafion/$60SiO_2$-$30P_2O_5$-$10ZrO_2$ membranes was examined using a Perkin-Elmer TGA6 model in nitrogen flow. The samples were first heated up to 80°C at the rate of 10°C/min, held there isothermally for 1 hour, then heated from 80 to 600°C at the rate of 10°C/min. DTA scans were obtained using a Perkin-Elmer DTA7 with samples under nitrogen flow and heated to 800°C at the rate of 10°C/min. An Amray 1200C scanning electron microscope (SEM) was used to study large-scale morphologies of these composites, both surface and cross sections. After visual SEM examination, EDX was used to study the distribution of elements across the film thickness. The intensity of the sulfur peak in the x-ray energy spectrum quantifies SO_3^- group population within a selected area and was adopted as an internal reference for the polymer matrix.

For the fuel cell experiments, the Pt/C fuel electrodes (ETEK Inc.) with a Pt loading of 0.4 mg/cm^2, were impregnated with 0.6 mg/cm^2 of Nafion (dry weight) by applying 12 mg/cm^2 of 5% Nafion solution with a brush. The electrode area was 5 cm^2. The membrane electrode assembly (MEA) was prepared by heating the electrode/membrane/electrode sandwich (active area of electrode was 5 cm^2) to 90°C for 1 minute in a Carver Hot-Press with no applied pressure, followed by increasing the temperature to 130°C for 1 minute with no applied pressure and finally hot-pressing the MEA at 130°C and 2 MPa for 1 minute. The MEA was positioned in a single cell test fixture, which was then installed in the fuel cell test station (Globetech Inc., GT-1000). The test station was equipped for the temperature-controlled humidification of the reactant gases (H$_2$, O$_2$ and air) and for the temperature control of the single cell. Flow rates of the gases were controlled using mass flow controllers. The total pressure of the gases was controlled using back-pressure regulators.

For the performance evaluation of the PEMFC, the single cell was fed with humidified H$_2$ and O$_2$ at atmospheric pressure (reactant gas and water vapor pressure equal to 1 atm) and the temperature of the H$_2$ and O$_2$ humidifiers and of the single cell was raised slowly to 90°C, 88°C and 80°C respectively. During this period, the external load was maintained at a constant value of 0.1 ohm, to reach an optimal hydration of the membrane using the water produced in the single cell. After the single cell had reached stable conditions (i.e. current density remained constant over time at a fixed potential), cyclic voltammograms were recorded at a sweep range of 20 mV s^{-1} and in the range of 0.1 V to 1 V vs. reversible hydrogen electrode (RHE) for one hour, in order to determine the electrochemically active surface area. Cell potential vs. current density measurements were then made under the desired conditions of temperature and pressure in the PEMFC. All PEMFC experiments were carried out at the cell temperatures of 80°C, 130°C and 140°C with the total pressure (reactant gas plus water vapor pressure) at 1 or 3 atm to maintain a relative humidity of 90-100%. The flow rates of gases were two times stoichiometric.

RESULTS AND DISCUSSION
Characterization of Membranes

The TGA curves for the original Nafion and Nafion/SiO$_2$-P$_2$O$_5$-ZrO$_2$ are shown in Figure 1. The first derivative indicates the temperatures at which the mass is changing fastest. From the figure is seen that the original Nafion membrane retains more than 95% of its weight up to 330°C. The infiltrated sample loses more weight because of the removal of organic groups and water from their gel component. The original Nafion membrane seems to degrade in three steps. The first derivative curve shows that these steps occur at 370, 450 and 515°C. This three-step decomposition has been reported in the literature within about ±4°C (12). The decomposition of

Nafion involves initial cleavage of the C-S bonds, leading to SO₂,·OH radicals, and a carbon-based radical which further degrades (13). The incorporation of sol-gel oxides does not seem to alter the thermal behavior of Nafion. However, the presence of the oxides contributes to higher weight loss at low temperature. The additional weight loss results from the removal of residual organic groups and water produced by condensation reactions. The derivative curve for the infiltrated Nafion membrane has a sharp peak centered at 350°C, indicating the elimination of organic groups, since this sample was treated only at 80°C before analysis.

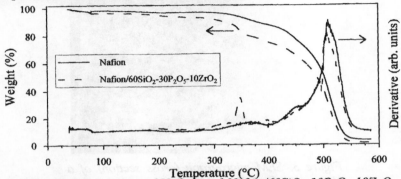

Figure 1. TGA curves of Nafion and Nafion/60SiO₂-30P₂O₅-10ZrO₂ membranes

Figure 2 shows the DTA curves for the Nafion and composite membrane treated at 150°C. The two of them are very similar with a very broad endothermic peak

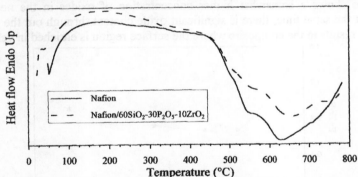

Figure 2. DTA curves of Nafion and Nafion/60SiO₂-30P₂O₅-10ZrO₂ membranes

between 100 and 450°C, and two exothermic peaks centered at about 550 and 625-650°C. The incorporation of the SiO₂-P₂O₅-ZrO₂ does little to alter the DTA behavior of Nafion, in part because the broadness of the observed peaks for Nafion overlap the temperature range for the characteristic evolution of the SiO_2-P_2O_5-ZrO_2 gel where removal of adsorbed water and organics, and elimination of hydroxyl groups occur.

Figure 3 is a SEM micrograph of a cross section of a Nafion/$60SiO_2$-$30P_2O_5$-$10ZrO_2$ sample, showing a thickness of 125 μm. The micrograph shows a uniform appearance without surface-attached layers, possibly with some particles of gel.

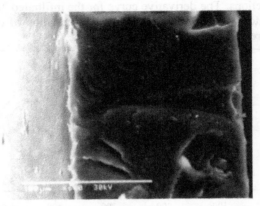

Figure 3. SEM micrograph (cross section) of a Nafion/$60SiO_2$-$30P_2O_5$-$10ZrO_2$ membrane

Elemental profile rating P+Zr:S is displayed in Figure 4 for an infiltrated sample. The phosphorus plus zirconium line (very close energies) was chosen in the interest of a better resolution because their combined intensity is greater than that of silicon. The profile shows a somewhat higher concentration of oxides in the near-surface regions. At the same time, there is significant oxide content through out the thickness. A gradient results in the composite where the surface region is enriched in gel.

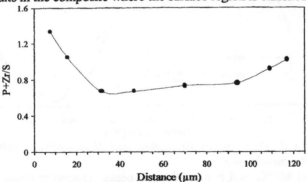

Figure 4. Elemental profile P+Zr:S of a Nafion/$60SiO_2$-$30P_2O_5$-$10ZrO_2$ membrane across the film thickness

Materials for Electrochemical Energy Conversion and Storage

Performance of Membranes in Fuel Cell Test

The main objective of this study was to determine whether or not the incorporation of SiO_2-P_2O_5-ZrO_2 via sol-gel processing in Nafion 115 could enhance the current density obtained at a fixed potential at a temperature of 130 to 140°C. By limiting the total pressure to 3 atm, the maximum operating temperature investigated was 140°C since the vapor pressure of water at this temperature is 3.5 atm.

Typical cyclic voltammograms for the cathode with the unmodified Nafion 115 and Nafion 115 composite were taken and by integrating the oxidation peak at 0.1 V vs. RHE and assuming a coulombic charge of 220 µC cm^{-2} for a smooth platinum surface, an average roughness factor of 135cm^{2}cm^{-2} for both the unmodified and composite membranes was obtained.

All cell potential (E) versus current density (i) data were analyzed by fitting the PEMFC data points to Equation 1:

$$E = E_o - b \log i - R i \qquad (1)$$

where E_o is the observed open cell potential, b is the Tafel slope and R accounts for the linear variation of overpotential with current density, primarily due to ohmic resistance. The exchange current density (i_o) of the oxygen reduction reaction can be calculated by using Equation 2:

$$E_o = E_r + b \log i_o \qquad (2)$$

where E_r is the theoretical open cell potential.

The polarization curve shown in Figure 5 compares the original Nafion membrane and the infiltrated Nafion membrane treated at 80°C with the H_2/O_2 humidification bottles at a temperature of 130°C and the single cell temperature set at 130 and 140°C with a total pressure of 3 atm. At both single cell temperatures, a significant improvement is seen in the water management of the composite vs. the control Nafion membrane. The higher current densities and lower resistances (Table I) of the composite membrane are attributed to the water retention characteristic of the SiO_2-P_2O_5-ZrO_2 gel.

Table I. Fuel cell performance

Membrane	R (Ω cm^2)		I (mA cm^{-2}) – 130°C		I (mA cm^{-2}) – 140°C	
	130°C	140°C	0.9V	0.4V	0.9V	0.4V
Nafion control	1.3	2.1	8	280	8	200
Composite 80°C	0.32	0.89	2	915	1	370
Composite 150°C	0.22	0.69	1	1275	---	430

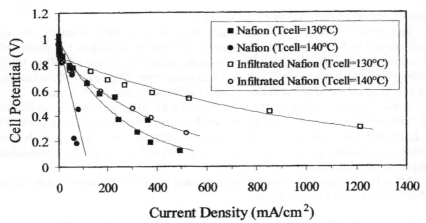

Figure 5. Cell potential vs. current density of Nafion and Nafion/60SiO₂-30P₂O₅-10ZrO₂ treated at 80°C.

Figure 6 (Table I) compares the two types of infiltrated Nafion membranes treated at 80 and 150°C. The infiltrated sample treated at 150°C demonstrates higher current densities and lower resistances at both cell temperatures than the sample treated at 80°C. The treatment at 150°C gives more complete removal of organics creating a porous material in the gel component with greater surface area and higher density of hydroxyl groups more capable of water retention, thereby increasing proton conductivity.

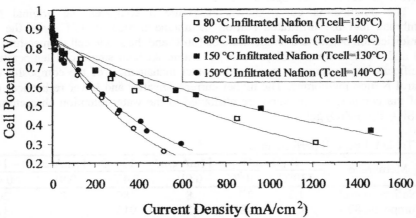

Figure 6. Cell potential vs. current density of Nafion/60SiO₂-30P₂O₅-10ZrO₂ treated at 80 and 150°C.

Figure 7 shows the two infiltrated Nafion samples with H_2/O_2 humidification bottles and single cell at 130°C with a total pressure of 3 atm, and the Nafion membrane operating at a single cell temperature of 80°C with hydrogen-oxygen humidification bottles at 90°C and 88°C respectively and 1 atm of pressure as the standard for fuel cell performance. The partial pressure of the reactant gases (H_2/O_2) under both operating conditions is ~ 0.5 atm due to the increase of vapor pressure from ~0.5 atm at 80°C to ~2.5 atm at 130°C. The polarization curves show an increase in operating temperature and lower resistances for infiltrated membranes providing superior PEMFC performance and temperature range than the present standard for PEMFC operation.

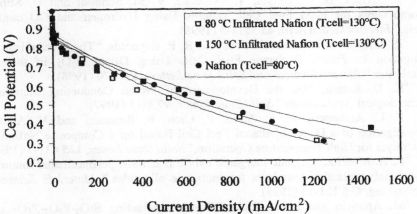

Figure 7. Cell potential vs. current density of Nafion and Nafion/60SiO₂-30P₂O₅-10ZrO₂ treated at 80 and 150°C.

CONCLUSIONS

Composite membranes were prepared by an impregnation of NAFION™ with $SiO_2-P_2O_5-ZrO_2$ sol. The incorporation of $SiO_2-P_2O_5-ZrO_2$ provided a significant increase in water retention capabilities for Nafion, allowing the composite membranes to be operated at elevated temperature. The infiltrated membrane treated at 150°C showed better performance, compared to a sample treated at 80°C, due to its higher hydroxyl group content. In addition, composite membranes were less susceptible to high temperature loss of proton conductivity than unmodified Nafion.

ACKNOWLEDGEMENTS

M. Aparicio wishes to thank the Spanish Ministry of Education and Culture for their financial support (grant PF99 0001828914).

REFERENCES

[1] M. Wakizoe, O. A. Velev and S. Srinivasan, "Analysis of Proton Exchange Membrane Fuel Cell Performance with Alternate Membranes," *Electrichimica Acta*, **40** 335-44 (1995).

[2] S. J. Lee, S. Murkerjee, E. A. Ticianelli and J. McBreen, "Electrocatalysis of CO Tolerance in Hydrogen Oxidation Reaction in PEM Fuel Cells," *Electrochimica Acta*, **44** 3283-93 (1999).

[3] M. C. Denis, G. Lalande, D. Guay, J. P. Dodelet and R. Schulz, "High Energy Ball-Milled Pt and Pt-Ru Catalysts for Polymer Electrolyte Fuel Cells and their Tolerance to CO," *Journal of Applied Electrochemistry*, **29** 951-60 (1999).

[4] J. Divisek, H. F., Oetjen, V. Peinecke, V. M. Schmidt and U. Stimming, "Components for PEM Fuel Cell Systems Using Hydrogen and CO containing Fuels," *Electrochimica Acta*, **43** 3811-5 (1998).

[5] R. J. Bellows, E. Marucchi-Soos and R. P. Reynolds, "The Mechanism of CO Mitigation in Proton Exchange Fuel Cells Using Dilute H_2O_2 in the Anode Humidifier," *Electrochemical and Solid-State Letters*, **1** 69-70 (1998).

[6] K. D. Kreuer, "On the Development of Proton Conducting Materials for Technological Applications," *Solid State Ionics*, **97** 1-15 (1997).

[7] P. L. Antonucci, A. S. Arico, P. Creti, E. Ramunni and V. Antonucci, "Investigation of a Direct Methanol Fuel Cell Based on a Composite Nafion/Silica Electrolyte for High Temperature Operation," *Solid State Ionics*, **125** 431-7 (1999).

[8] K. A. Mauritz, "Organic-inorganic hybrid materials: perfluorinated ionomers as sol-gel polymerization templates for inorganic alkoxides," *Materials Science and Engineering*, **C 6** 121-33 (1998).

[9] M. Aparicio and L. C. Klein, "Proton conducting SiO_2-P_2O_5-ZrO_2 sol-gel glasses", *Ceramic Transactions* Vol. 123: Sol-gel Processing of Advanced materials: Commercialization and Applications, ed. X. Feng, E. J. A. Pope, S. Sakka, L. C. Klein and S. Komarneni, American Ceramic Soc., Westerville, OH, 2000.

[10] M. Nogami, K. Miyamura and Y. Abe, "Fast Protonic Conductors of Water-Containing P_2O_5-ZrO_2-SiO_2 Glasses," *Journal of the Electrochemical Society*, **144** [6] 2175-8 (1997).

[11] M. Nogami, R. Nagao, W. Cong and Y. Abe, "Role of Water on Fast Proton Conduction in Sol-Gel Glasses," *Journal of Sol-Gel Science and Technology*, **13** 933-6 (1998).

[12] I. D. Stefanithis and K. A. Mauritz, "Microstructural Evolution of a Silicon Oxide Phase in a Perfluorosulfonic Acid Ionomer by an in Situ Sol-Gel Reaction. 3. Thermal Analysis Studies," *Macromolecules*, **23** 2397-402 (1990).

[13] C. A. Wilkie, J. R. Thomsen and M. L. Mittleman, "Interaction of Poly(Methyl Methacrylate) and Nafions," *Journal of Applied Polymer Science*, **42** 901-9 (1991).

STUDY OF GLASS/METAL INTERFACES UNDER AN ELECTRIC FIELD: LOW TEMPERATURE/HIGH VOLTAGE

M. A. Alvarez*, and L. C. Klein
RUTGERS, The State University of New Jersey
Ceramic & Materials Engineering
607 Taylor Rd
Piscataway, NJ 08854-8065

ABSTRACT

In order to identify conditions leading to failure in systems that require an alkali ion conductor or alkali-containing component, we have studied an aluminum/binary sol-gel system. Dielectric properties before and after exposure to an electric field have been compared in lithium silicate gels containing 5, 10 and 15 wt % lithium oxide.

INTRODUCTION

A common problem in devices or systems incorporating an ion conductor is the degradation of the device at the contacts between the various components. Typically, the systems show degradation at glass/metal interfaces due to thermal cycling and aging [1]. The cause of degradation may vary from system to system, but in all cases, the presence of an electric field is a contributing factor. Systems that depend on an ion conductor include batteries, fuel cells, oxygen generators and sensors [2]. Different devices operate either at low temperature or at high temperature. The temperature regime depends on the value and the type (cation/anion) of ionic conductivity. That is the ionic conductor may conduct through defects, as in zirconia, or through mobile alkali ions, as in lithium electrolytes in lithium batteries [3].

In glasses containing alkali oxides, the current is carried almost exclusively by alkali ions. The mobility of these ions is much greater than the mobility of the network-forming ions at all temperatures. It is the concentration and mobility of the alkali ions that determines conduction characteristics [4]. In glasses, there is not a single value for the energy barrier between ion positions. There are often adjacent low-energy positions with a small energy barrier between them. Large barriers occur occasionally in accordance with the random nature of the glass structure.

In many cases, alkali silicates have been used as electrolytes. For that reason, ionic conductivity is the relevant property, which is measured using complex impedance spectroscopy under AC conditions [5]. In a few cases, dielectric properties have been measured as a function of frequency or relative humidity [6-9]. In the case of alkali silicates subjected to high field at room temperature, the question is whether or not there is significant alkali migration. Similarly we need to determine the effect of migration, if it occurs, on dielectric properties.

While ion migration in alkali-containing glasses and glass-ceramics has been studied for many years [10,11], the consequences of ion migration in operating devices have not received proper attention. Therefore, in this study, we have tried to simulate high field conditions that lead to ion migration below the glass transition temperature (T_g), through the use of an external electric field. High field conditions might be present inside these devices, although in operation the electric field may not be the only factor responsible for the degradation of the interface. The use of these devices is normally accompanied by the presence of heat, but our conditions have been room temperature only [12]. Subsequently, we have measured dielectric properties as a function of exposure to high field for various periods of time.

We have used the sol-gel process to prepare lithium-containing silicates, because it is a versatile technique for fabrication of amorphous materials [13]. In the past, we found this processing technique particularly useful in producing porous materials with high surface area and low density, which were desirable characteristics for electrolyte materials [14-17].

EXPERIMENTAL

Samples of lithium silicate were created using the sol-gel method [13]. A sample formulation of 5 wt % Li_2O 95 wt% SiO_2 is:

Table I: Sample formulation of 5%Li sol-gels

	Formula	Composition
Si	$Si(OC_2H_5)_4$	27.09ml
Li	$LiNO_3$	0.146ml
EtOH	Same as TEOS	
D-H_2O	4 mol times of TEOS = 64 ml	
H_2SO_4	1 ml	

The $LiNO_3$ was first mixed with sulfuric acid and ethanol. After mixing for 10 minutes, de-ionized water was added. After another 10 minutes, the TEOS was added. After mixing for 30 minutes, the samples were left to dry in an oven at 50°C for one week. In addition, samples with 10 and 15 wt % Li_2O were prepared. Samples are designated LS5, LS10 and LS15 to indicate lithium silicates with the indicated weight % of lithium oxide.

The samples varied in sizes, but they were generally rod-shaped or discs (diameter ~15mm, length ~10 mm). Upon drying, the presence of cracks was noticeable. The appearance was translucent and without coloration. The appearance did not change upon exposure to an Electro-Technic (BD-10A) High Frequency Generator. This generator produced an electric field of about 50 kV/inch (20,000 V/cm) at a frequency of 0.5 Megahertz. The samples were exposed to the electric field for different periods of time (0,1,2,3,4, and 5 minutes). Initially, upon exposure it was noticed that the sample surfaces suffered from the intense voltaic arc that this generator created. Therefore, the surface of the samples was protected from the voltaic arc by using silver paint. Exposed samples were characterized by the use of x-ray diffraction analysis, thermal analysis and chemical analysis.

A GenRad RLC Digibridge was used to make the dielectric measurements. This instrument measures capacitance on disc-shaped samples. Following calibration with a penny and standard capacitors, using the tweezers sample holder, the exposed gel samples were measured.

Ion Coupled Plasma (Dionex, ICP) spectroscopy was used to measure cation concentration in solutions. Solutions were prepared by dissolving the samples (10 mg) in hydrofluoric acid (1 ml), which was diluted with water to prevent damage to the detector, similar to our earlier work [18]. The samples were labeled according to their position with respect to the anode and cathode.

RESULTS AND DISCUSSION

Our hypothesis is that the electric field causes ion migration during the exposure time. While the field is present, alkali ions respond to the high voltage, even at room temperature. When the field is removed, the ions are not mobile, so they remain in a non-uniform distribution and are unable to relax to a homogeneous composition.

X-ray diffraction (XRD) patterns of fresh and exposed samples showed no differences with exposure time therefore are not exposed in this paper. Furthermore, the x-ray spectra showed x-ray amorphous behavior with no crystallization. Thermal analysis, both calorimetry and thermogravimetry, confirmed x-ray characterization, showing a behavior with little difference among samples. This lack of a difference also suggested that the samples did not heat up during exposure or experience drying due to a rise in temperature. All samples showed weight loss upon heating due to the removal of water(~100°C) and removal of solvents (~150-250°C). The higher temperature weight loss (~400-700°C) was attributed to removal of surface hydroxyl groups.

Dielectric measurements were used as an indication of the effect of high voltage on the chemical composition of the samples, under the assumption that once the alkali ions respond to the high voltage, they cannot relax at room temperature. When a dielectric is inserted between the plates of a capacitor, the capacity for charge storage at constant voltage is increased. The capacitance depends on sample thickness, sample area, and composition. The increase in capacitance is proportional to the permittivity of the dielectric, making it proportional to the dielectric constant of the material. The increase in capacitance results from the polarization of the dielectric. In the case of alkali silicates, the alkali ions contribute to the ionic polarization. We are assuming that the exposure to high voltage alters the samples in such a way that the dielectric properties will reflect the resulting ionic polarization.

Electrical measurements included capacitance, dissipation factor, and series resistance. The dielectric constant was calculated.

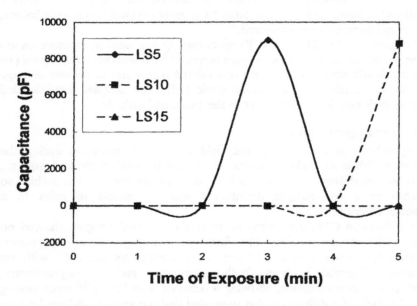

Figure 1: Normalized Capacitance vs. Time-of-Exposure

In Figure 1, the measured capacitance is plotted as a function of time exposed to the electric field. While all samples of a given composition were from the same batch, the measured capacitance is not the same. Therefore, the capacitance for the exposed sample is normalized to the capacitance for the fresh sample. The effect of normalizing the values is to exaggerate the effect of exposure time and make it easier to compare compositions. For the LS5 the trend seemed to increase to a maximum around 3 minutes of exposure. In the case of the LS10, there was little change until 5 minutes. The LS15 samples did not show any response as a capacitor. The behavior for LS15 is attributed to the large amount of lithium present, which would make this composition a better conductor and therefore minimize its ability to act as a capacitor. The increase in charge carriers would have decreased the charge storage capacity of the glass.

Dissipation factor indicates the retardation angle (current leads the voltage), which causes a dissipation of electrical energy. This energy can heat up the dielectric and cause circuit imbalance or total breakdown if the losses are too great. Figure 2 shows the dissipation factor behavior for the three samples studied. For LS5, there is little change, so little energy is dissipated. As the lithium content increases, there is an increase for the value of dissipation factor. This fact is attributed to the motion of charge carriers. This effect is most noticeable at the beginning (1 min) of the exposure and then the value decreases, indicating that the sample relaxes after its initial response.

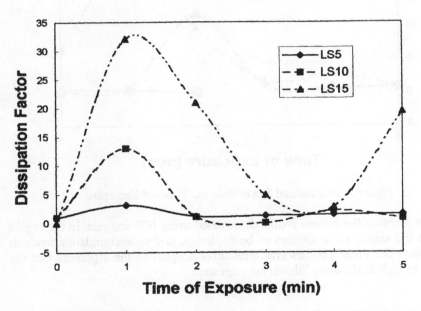

Figure 2: Normalized Dissipation Factor vs. Time-of-Exposure

According to Ohm's Law, the ratio of the voltage to the current is the resistivity. Since the current is based on the flow of electrons, then anything that impedes their movement would cause the resistance to increase. Figure 3 shows the resistance behavior of the samples. LS5 showed little change, although there was some increase in resistance with time of exposure. LS10 showed a spike in resistance at around 3 minutes, while LS15 showed a spike at around 4 minutes. These spikes were attributed to a lithium depletion or migration in the direction of the field. Since the Li follows the electric field, we speculate that there are fewer charge carriers at the anode side. This reduction of carriers raises the resistance of the material since the current is not able to flow.

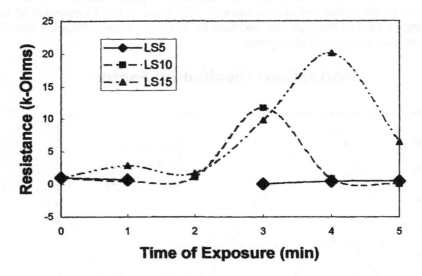

Figure 3: Normalized Resistance vs. Time-of-Exposure

Figure 4 shows the lithium profile determined using ICP analysis. In the region closest to the anode, there appears to be depletion, and an accumulation towards the cathode side. These profiles give qualitative support to the argument that the exposure to high field causes lithium ion migration.

Materials for Electrochemical Energy Conversion and Storage

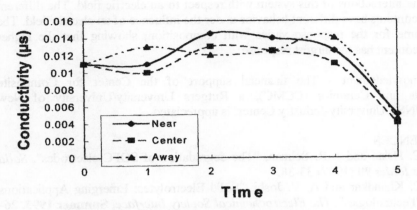

ICP Li Time

Figure 4: Spatial Profile of Lithium in LS5

Under alternating current, we assume that alkali ions are not responding to the AC field in an ideal way. Rather, the dielectric dissipates some energy, which contributes to dielectric loss. Therefore, the dielectric measurements were recorded at different frequencies. The capacitance and the dissipation factor were measured and the dielectric constant was calculated. The dielectric constant increased from about 5 for LS5 to 7 for LS10 to 13 for LS15. The dielectric constant increased with exposure in LS10 and LS15, but was not changed in LS5.

The time dependence, as well as the composition dependence, of these trends can be modeled to find a characteristic relaxation behavior. Assuming that a relaxation time can be calculated, a diffusion coefficient that is representative of this room temperature process can be extracted. The reason for searching for a diffusion coefficient is that a diffusion model is a reasonable assumption for the process going on during exposure to high voltage. Having order of magnitude estimates for this process would make it possible to make some predictions about the effects of exposure over long periods of time.

A complicating factor at room temperature is the influence of adsorbed moisture. Samples were kept in an oven at a temperature of 100°C to prevent water absorption, but during measurements the samples were exposed briefly to atmospheric moisture. In past studies, significant proton conductivity has masked ionic conductivity. It may be necessary to measure samples as a function of temperature, but the danger is that temperature will remove the effects of the high voltage.

CONCLUSION

Dielectric analysis of lithium silicate gels has provided interesting information about the interactions of this system with respect to an electric field. The different lithium contents have different behavior under the influence of an electric field. The time frame for the response varies with composition, showing that the higher lithium content has the largest response.

Acknowledgment - The financial support of the Center for Composite Materials and Ceramics (CCMC), a Rutgers University/University of New Mexico/NSF University-Industry Center, is appreciated.

REFERENCES

1. Y. C. Hsiao and J. R. Selman, "The degradation of SOFC electrodes", *Solid State Ionics* **98** (1997) 33-38.
2. A. C. Khandkar and A. V. Joshi, "Solid Electrolytes: Emerging Applications and Technologies", *The Electrochemical Society Interface*, **Summer 1993**, 26-33.
3. H. L. Tuller, "Ionic conduction in nanocrystalline materials", *Solid State Ionics* **131** (2000) 143-157.
4. M. Tomozawa, J. F. Cordaro and M. Singh, "Applicability of weak electrolyte theory to glasses", *J. Non-Crystal. Solids* **40** (1980) 189-196.
5. L. C. Klein, S-F. Ho, S-P. Szu and M. Greenblatt, "Applications of AC complex impedance spectroscopy to understanding transport properties in lithium silicate gels," in *Applications of Analytical Techniques to the Characterization of Materials*, Ed. D. L. Perry, Plenum, NY, 1991, pp. 101-118.
6. W. Cao, R. Gerhardt and J. B. Wachtman, "Low permittivity porous silica by a colloidal processing method". in *Advances in Ceramics* **Vol. 26**: Ceramic Substrates and Packages for Electronic Applications, ed. M. F. Yan, American Ceramic Society, Westerville, OH, 1989.
7. S. K. Saha and D. Chakravorty, "Conductivity relaxation in sol-gel derived Glasses", *J. Phys. D: Appl. Phys.* **23** (1990) 1202-1206.
8. M. A. Villegas, J. R. Jurado and J. M. Fernandez Navarro, "Dielectric properties of R_2O-SiO_2 glasses prepared via sol-gel", *J. Materials Science* **24** (1989) 2884-2890.
9. L. C. Klein and S-F. Ho, "Lithium ion conducting silicate gels: synthesis and characterization," *Transactions 20: Glasses for Electronic Applications*, Ed. K. M. Nair, Am. Ceram. Soc., 1991, pp. 221-233.
10. J. Lau and P. W. McMillan, "Interaction of Sodium with simple glasses, Part 1 Vitreous Silica", *J. Materials Science* **19** (1984) 881-889.

11. R. J. Araujo and F. P. Fehler, "Sodium Redistribution between oxides Phases", *Journal of Non-Crystal. Solids* **197** (1996) 154-163.
12. M. Wasiucionek and M. W. Breiter, "Lithium conduction in inorganic gel monoliths containing LiNO3 or LiBr at low temperatures", *Solid State Ionics* **119** (1999) 211-216.
13. L. C. Klein, "Sol-gel processing of ionic conductors," *Solid State Ionics,* **32/33** (1989) 639-645.
14. L. C. Klein, E. Mouchon, V. Picard and M. Greenblatt, "Sol-gel lithium silicate electrolyte thin films," MRS Vol. 346: *Better Ceramics through Chemistry VI,* eds. C. Sanchez, C. J. Brinker, M. L. Mecartney and A. Cheetham, Materials Research Society, Pittsburgh, PA, 1994, pp. 189-200.
15. S-P. Szu, M. Greenblatt, and L. C. Klein, "The effect of precursors on the ionic conductivity in lithium silicate gels," *Solid State Ionics* **46** (1991) 291-297.
16. L. C. Klein, H. Wakamatsu, S-P. Szu and M. Greenblatt, "Effect of lithium salts on the ionic conductivity of lithium silicate gels," *J. Non-Crystal. Solids* **147&148** (1992) 668-671.
17. L. Laby, L. C. Klein, J. Yan and M. Greenblatt, "Ionic conductivity of lithium aluminosilicate xerogels and thin films", *Solid State Ionics* **81** (1995) 217-224.
18. N. Le Bars and L. C. Klein, "Lithia distribution in infiltrated silica gels," *J. Non-Cryst. Solids* **122** (1990) 291-297.

Lithium–Ion Batteries

OLIVINE-TYPE CATHODES FOR LITHIUM BATTERIES

A. Yamada,* M. Hosoya, S. C. Chung, and K. Hinokuma
Frontier Science Laboratories, Sony Corporation
2-1-1 Shinsakuragaoka, Hodogaya, Yokohama 240-0036, Japan

Y. Kudo and K.Y. Liu
Technical Support Center, Sony Corporation
4-16-1 Okada, Atsugi 243-0021, Japan

ABSTRACT

The reaction mechanism of Olivine-type cathodes, $Li_x(Mn_yFe_{1-y})PO_4$ ($0<x,y<1$), was investigated by x-ray diffraction, Mössbauer spectroscopy, OCV measurements, x-ray absorption spectroscopy, and ab initio calculation. The phase diagram in (x,y) plane was clarified in terms of (1) the orthorhombic (S. G. Pmnb) unit-cell dimensions, (2) valence states of Mn and Fe, and (3) single phase - two phase reaction forms. The strong electron ($Mn^{3+}:3d^4$) - lattice interaction in the charged state will be highlighted as the intrinsic obstacle to generate full theoretical capacity, ca. 170mAh/g, of the Mn-rich ($y>0.7$) phase, followed by the promising cathode performance of the optimized electrode-powders of $LiFePO_4$ and $Li(Mn_{0.6}Fe_{0.4})PO_4$.

INTRODUCTION

Since the demonstration of reversible electrochemical lithium insertion-extraction for $LiFePO_4$ in 1997[1], lithium transition metal phosphates with ordered-olivine structure, $LiMPO_4$ (M=Co, Ni, Mn, Fe, Cu), have attracted much attention as promising new cathode materials for rechargeable lithium batteries[2-7]. The initially included lithium ion Li^+ per transition metal ion M^{2+} can be extracted in the first charge process, compensating for the oxidation of M^{2+} to M^{3+}, and transferred to the carbon anode through the non-aqueous electrolyte.

$$\text{Cathode:} \quad LiM^{2+}PO_4 \quad \leftrightarrow \quad M^{3+}PO_4 + Li^+ + e^-$$
$$\text{Anode:} \quad C_6 + Li^+ + e^- \quad \leftrightarrow \quad C_6Li$$

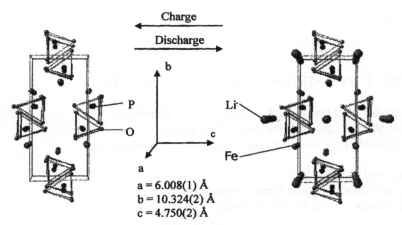

Fig. 1 The crystal structure of LiFePO₄

The opposite movement of lithium ions and electrons occurs in the discharge process, while the transition metal M is reduced from trivalent to divalent. LiMPO$_4$ crystal has an orthorhombic unit cell (D_{2h}^{16} – S. G. Pmnb), which accommodates four units of LiMPO$_4$[1-3]. As a typical example, LiFePO$_4$ has an unit-cell dimensions of a=6.008(1)Å, b=10.324(2)Å, and c=4.694(1)Å[1,2,4]. The visual expressions of the crystal structure is given in Fig. 1.

The charge-discharge reaction of presently used materials such as layered rock salt systems, LiCoO$_2$, LiNiO$_2$ (S. G.: R3m), and spinel framework system Li$_{0.5}$MnO$_2$ (S. G.: Fd3m) are all based on the M^{4+}/M^{3+} couple in the edge-sheared MO$_6$ octahedra in the closed-packed oxygen array and generates ca. 4V.

$$LiM^{3+}O_2 \quad \leftrightarrow \quad M^{4+}O_2 + Li^+ + e^-$$

Therefore, the presence of large tetrahedral polyanion (PO$_4$)$^{3-}$ and the use of the M^{3+}/M^{2+} redox couple are distinctive features of the olivine-type cathodes. The P$_{tet}$-O-M$_{oct}$ linkage in the structure induces the superexchange interaction that tunes the M^{3+}/M^{2+} redox energy to useful levels (3.4V, 4.1V, and 4.8V for Fe^{3+}/Fe^{2+}, Mn^{3+}/Mn^{2+}, and Co^{3+}/Co^{2+}, respectively)[1]. The strong P$_{tet}$-O covalency stabilizes the antibonding Fe^{3+}/Fe^{2+} (3d$^6\pi$*) state of primarily cationic origin to generate an appropriately high voltage, which is called the "inductive effect"[1]. The stable nature of the olivine-type structure having a (PO$_4$)$^{3-}$ polyanion with a strong P-O covalent bond provides not only an excellent cycle-life but also a safe system when the battery is fully charged; the reactivity is low in the combustion reaction with the organic electrolyte[4]. The energy density of olivine-type LiMPO$_4$ is equal to that of presently used materials, based on the theoretical charge-discharge capacity of 170mAh/g obtained from M^{3+}/M^{2+} one-electron redox reaction of Li_xMPO_4 $(0 \leq x \leq 1)$[4], and appropriately high voltage of >3.4V.

In the olivine-type $LiMPO_4$ family, $LiFePO_4$, $LiMnPO_4$, and their solid solution system, $Li(Mn_yFe_{1-y})PO_4$ [1], look promising because they operate at 3.4-4.1V vs. Li/Li^+, which is providential because it is not so high as to decompose the organic electrolyte but is not so low as to sacrifice energy density. The positions of redox couples were also confirmed by the first-principle calculation[4]. The use of $LiMnPO_4$ is of particular interest because the position of the Mn^{3+}/Mn^{2+} couple, 4.1V versus Li/Li^+ [1], is compatible with present lithium-ion batteries and generates high energy density. However, it has been shown that the capacity at 4.1V is not achieved without Fe coexisting with Mn at the octahedral 4c site[1]. Padhi et al. have performed systematic experiments on the electrochemical charge and discharge characteristics of $Li(Mn^{2+}_yFe^{2+}_{1-y})PO_4$ (y = 0.25, 0.50, 0.75, 1.0) and reported that the width of the 4.1V plateau (Mn^{3+}/Mn^{2+}) relative to that of the 3.4V plateau (Fe^{3+}/Fe^{2+}) increases as Mn content y is increased, but that the total capacity rapidly decreases at y>0.75 [1].

In this paper, the charge-discharge reaction mechanism of Olivine-type cathodes, $Li_x(Mn_yFe_{1-y})PO_4$ $(0 \leq x, y \leq 1)$, was investigated in detail using various technique. The phase diagram in (x,y) two-dimensional plane was clarified in terms of (1) the orthorhombic (D_{4h}^{12}: Pmnb) lattice constants, (2) the valence states of Mn and Fe, and (3) single phase – two phase reaction forms. The strong electron (Mn^{3+}:$3d^4$-$e_g\sigma^*$) - lattice interaction (Jahn-Teller effect) in the charged state is highlighted as main obstacle to generate the full theoretical capacity of the Mn-rich (y>0.7) phase, and the impressive cathode performance of $LiFePO_4$ and $Li(Mn_{0.6}Fe_{0.4})PO_4$ will be demonstrated. The essential strategies for a design of practical olivine-type cathode materials will be discussed, particularly on the importance of the powder engineering of the electrode composites.

METHODOLOGY
Experimental

The $Li(Mn_yFe_{1-y})PO_4$ (6 samples: y = 0, 0.2, 0.4, 0.6, 0.8, 1.0) compounds was prepared by solid-state reaction of $FeC_2O_4 \cdot 2H_2O$, $MnCO_3$, $NH_4H_2PO_4$, and Li_2CO_3. They were dispersed into acetone, then thoroughly mixed and reground by high-energy ball milling. After evaporating the acetone, the olivine phase was synthesized in purified N_2 gas flow (800cc/min) to prevent the formation of trivalent compounds as impurities. The mixture was first decomposed at 280°C for 3h to disperse the gases and firmly reground again by ball milling, then sintered for 24 hours at 600°C. In either of the above ball-milling processes, carbon black as conducting additives can be mixed together for homogeneous electrodes. This was possible because of no oxidation reaction during sintering in inert gas, and led to a dense conductive electrode, higher materials utilization for redox reaction, and better high-rate performance.

Chemical oxidation to obtain $(Mn_yFe_{1-y})PO_4$ was performed by reacting $Li(Mn_yFe_{1-y})PO_4$ with nitronium tetrafluoroborate (NO_2BF_4) in acetonitrile[5-7]. The redox potential of NO_2^+/NO_2 is ca. 5.1V vs. Li/Li^+ and is effective to oxidize $Li(Mn_yFe_{1-y})PO_4$ with redox potentials of 3.4V (Fe^{3+}/Fe^{2+}) and 4.1V (Mn^{3+}/Mn^{2+}) vs. Li/Li^+. The reaction is written as

$$Li(Mn_yFe_{1-y})PO_4 + NO_2BF_4 \rightarrow (Mn_yFe_{1-y})PO_4 + LiBF_4 + NO_2.$$

To ensure complete reaction, the amount of NO_2BF_4 added was double the amount estimated from reaction (1). After 8.5g of NO_2BF_4 was dissolved into 300ml acetonitrile, 5g of $Li(Mn_yFe_{1-y})PO_4$ was added and the mixture was stirred for 24 hours at room temperature under bubbling purified Ar gas. The product was filtered and washed several times with acetonitrile to remove impurities before it was dried in vacuum at 70°C. Lithiated samples of $Li_x(Mn_yFe_{1-y})PO_4$ $(0<x<1)$ were prepared by reacting $(Mn_yFe_{1-y})PO_4$ with various amounts of LiI (High Purity Chemicals, 99.9%) in acetonitrile. The hygroscopic LiI powder was treated in a dry atmosphere and the ratio of acetonitrile to $(Mn_yFe_{1-y})PO_4$ was set at 200ml to 0.2g. The solution was stirred for 24 hours at room temperature and the product was filtered and washed several times with acetonitrile to remove impurities before it was dried in vacuum at 70°C. The compositional analysis by ICP-AES revealed that the amount of lithium after the reactions was close to the expected values.

X-ray powder diffraction (XRD, RINT-2500v, Rigaku Co.) with Cu-K_α radiation was used to identify the phases and analyze the structure. The lattice constants were calculated using the least square method with a Si standard. The diffraction profiles were measured in the slow-scan mode (0.5°/min). The ^{57}Fe Mössbauer spectra in transmission geometry were collected with a ^{57}Co γ-ray source. Velocity calibration was made using the data of α-Fe at room temperature. The sample thickness was adjusted so that the Fe content was ca. 8mg/cm^2. X-ray absorption measurements were performed at the Industrial Consortium Beamline BL16B2 in SPring-8 using synchrotron radiation from the electron storage ring at an electron energy of 8GeV.

The performance of the $Li(Mn_yFe_{1-y})PO_4$ cathodes was evaluated using a coin-type cell (Size 2025) with a lithium metal anode. The cathode was a mixture of $Li(Mn_yFe_{1-y})PO_4$/ carbon black / polyvinylidene fluoride (PVDF) with weight ratio 90/8/2 (60mg of total weight), and the electrolyte was a 1M LiPF$_6$ – propylene carbonate/dimethyl carbonate (PC/DMC) solution. The galvanostatic charge-discharge experiment with 0.12mA/cm^2 was performed between 2.0V and

4.5V at 23°C. At the end of the charge process, the cell voltage was kept constant at 4.5V until the current density decreased to <0.012mA/cm^2.

First-Principle Calculations

Ab initio calculation was applied to consider the electronic structures, operating voltages, and the Mössbauer parameters[4,5]. All calculations were done within the density functional theory approximation using the gradient corrected functional for exchange and correlation. The FP-LAPW code Wien97 was used. In the present calculation, l_{max} = 10 was used as the upper limit for the angular expansions of the wave functions inside the muffin-tin spheres and the energy cutoff for the plane-wave expansions in the interstitial region was 36.0 Ry (RMT*K = 7.2). The local orbital was added to the basis of iron in order to effect maximum flexibility. All calculations were conducted in a spin-polarized fashion. For Brillouin zone integration, 126 k points were used in the sampling with the exception of lithium metal where 3000 k points were employed

RESULTS AND DISCUSSIONS

Phase Diagrams

Shown in Fig. 2 are variations in the three lattice constants of the orthorhombic Pmnb lattice a, b, and c with lithium composition x. A two-phase reaction in Li$_x$FePO$_4$ (0≤x≤1) was apparent[1]. Then, the Mn substitution for Fe in the octahedral 4c sites results in (1) the formation of a two-phase Mn^{3+}/Mn^{2+} region at x<y (4.1V vs. Li/Li$^+$), where all of the three orthorhombic lattice constants show discrete elongation by Li insertion, and (2) a partial conversion of the reaction form for Fe^{3+}/Fe^{2+} region of x>y (3.4V vs. Li/Li$^+$) from two-phase to single-phase, where the a- and b- lattice constants increase and the c- lattice constant decreases[6,7]. The single-phase regions are shaded in Fig. 2 for an aid to understand the general tendency. As a result of this two-phase to single-phase conversion in Fe^{3+}/Fe^{2+} and the growth of the Mn^{3+}/Mn^{2+} region, the two-phase region in Fe^{3+}/Fe^{2+} disappears at y ≥0.6.

The (x, y) two-dimensional phase diagram of the Li$_x$(Mn$_y$Fe$_{1-y}$)PO$_4$ (0≤x,y≤1) system[7] is given as Fig. 3. Information is simply given on the single-phase or two-phase matter together with the valence states of Mn and Fe, and this establishes some general trends for the phase change. The (x, y) phase map in Fig. 3 is divided into four areas: (a) the unstable region close to the point (x, y) = (1, 0), (b) the two-phase region caused by Mn^{3+}/Mn^{2+} (y≥x), (d) the two-phase region caused by Fe^{3+}/Fe^{2+} (a part of y≤x), and (c) the single-phase region caused by Fe^{3+}/Fe^{2+} connecting (b) and (d). The change in the valence state of Fe was analyzed using Mössbauer spectroscopy[6], and showed the Mn^{3+}/Mn^{2+} and Fe^{3+}/Fe^{2+} redox reactions in the compositional regions x≤y and x≥y, respectively. The boundaries at x=y were distinct.

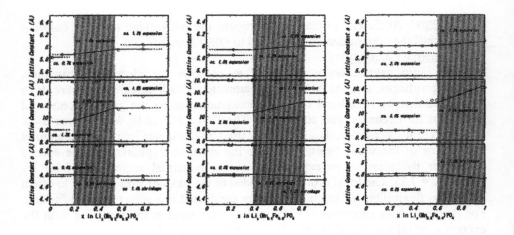

Fig. 2 Variations in the three lattice constants as a function of Li content x

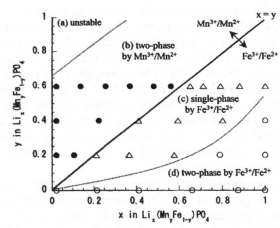

Fig. 3 The (x, y) two-dimensional phase diagram of $Li_x(Mn_yFe_{1-y})PO_4$

It is noteworthy that the Mn^{3+}/Mn^{2+} redox reaction (region of $y \geq x$) entirely proceeds in a two-phase manner. This can be readily understood by the first-order transition between the Jahn-Teller-active Mn^{3+} phase with cooperative elastic deformation and the Jahn-Teller-inactive Mn^{2+} phase. Another important aspect is that the single-phase region (c) appears only in Fe^{3+}/Fe^{2+} in a Mn/Fe solid solution system but not in Li_xFePO_4. The origin of the appearance of this single-phase region (c) is not clear, but it is reasonable to speculate conceptually that the random distribution of divalent manganese in region (c) may dilute the weak cooperative interaction that discretely adjusts the framework to lithium insertion/extraction and makes the inherent Fe^{3+}/Fe^{2+} redox reaction in Li_xFePO_4 ($0 \leq x \leq 1$) a two-phase type[1].

The most important finding here is the unstable region (a)[5,7], and this causes the sudden capacity drop with large manganese content[1]. We could not isolate the material in this composition range (a) in an equilibrium condition[5]. In order to clarify what causes this instability, the unit cell dimensions were measured along two scales, charged state (x=0) and discharged state (x=1), as a function of the Mn content. The former is trivalent line and the latter is divalent line[5].

Jahn-Teller Instability by Mn^{3+}

Figure 4 summarizes the variation of lattice constants as a function of Mn content y. For all choices of Mn content y, oxidation of $Li(Mn^{2+}_yFe^{2+}_{1-y})PO_4$ to $(Mn^{3+}_yFe^{3+}_{1-y})PO_4$ induces elongation of the c-axis and shrinkage in the a- and b-axes[5]. A simple isotropic expansion of the orthorhombic lattice in the series of $Li(Mn^{2+}_yFe^{2+}_{1-y})PO_4$ is consistent with the previous reports, as well as the values of the three lattice constants[1]. On the other hand, a characteristic feature was observed for the series of $(Mn^{3+}_yFe^{3+}_{1-y})PO_4$: Mn^{3+} in 4c sites induced elongation of the a-axis and shrinkage in the b- and c-axis, although the change in the c-axis is slight[5]. Note that the dashed-lines are used for the unstable region (a).

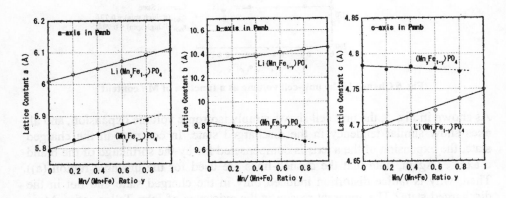

Fig. 4 Variation of the orthorhombic lattice constants as a function of Mn content y in $Li(Mn^{2+}_yFe^{2+}_{1-y})PO_4$ (open circles) and $(Mn^{3+}_yFe^{3+}_{1-y})PO_4$ (closed circles).

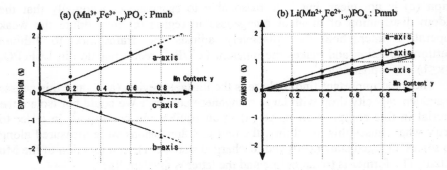

Fig. 5 The percent of unit-cell distortions as a function of Mn content for (a) $(Mn^{3+}_y Fe^{3+}_{1-y})PO_4$ (charged state) and (b) $Li(Mn^{2+}_y Fe^{2+}_{1-y})PO_4$ (discharged state).

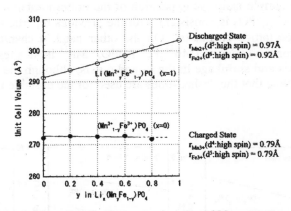

Fig. 6 Change in the unit-cell volume as a function of Mn content y

As shown in Fig. 5, the unit cell shows simply isotropic volume expansion, and all axes show similar tendency in the discharged state. In contrast, in the charged state, the expansion of the a axis is compensated for by the shrinkage of the b and c axes[5]. (Again, note that the dashed-lines are used for the unstable region (a)) Then, why is lattice distortion induced only in the charged state and not in the discharged state? The apparent reason is the existence of Jahn-Teller-active Mn^{3+} in the charged state[8].

The monotonic increase of the unit-cell volume is observed in a series of $Li(Mn^{2+}_y Fe^{2+}_{1-y})PO_4$, while the volume is almost constant in a series of $(Mn^{3+}_y Fe^{3+}_{1-y})PO_4$. (Fig. 6) This tendency can be explained by the difference of Pauling's ionic radius for the high-spin configuration as denoted in Fig. 6.

Electronic Structures

High-spin configurations in the strongly distorted oxygen octahedra in an olivine-type structure are shown below, where ab-initio calculation was performed for a better understanding of the electronic structure in olivine-type materials[5]. The total density of states (DOS) for $LiFePO_4$ is shown in Fig. 7, left. The narrow band near the Fermi level can be assigned to the 3d band of Fe. The so-called antibonding e_g and nonbonding t_{2g} states under the O_h symmetry can be identified despite the degeneracy induced by the C_s symmetry of the Fe site (Fig. 7, right). The spin-up 3d states are completely filled, while there is a peak at 0.8 eV below the Fermi level for the spin-down states and the rest are unfilled. Iron atoms in $LiFePO_4$ are in the +2 oxidation state, with a formal $3d^6$ electronic configuration. In a high-spin configuration, one will expect a magnetic moment of 4 μ_B per Fe atom. We have integrated the charge density in the unit cell and found that there is a magnetic moment of 3.8 μ_B per $LiFePO_4$ unit. This together with the DOS shows that the 3d state of Fe is high spin. We have also performed the calculation for $FePO_4$ and $LiMnPO_4$ and confirmed the high-spin configuration[5].

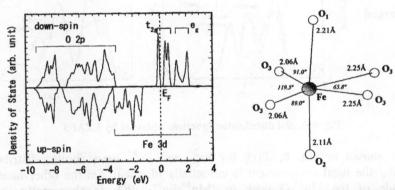

Fig. 7 Total density of states of $LiFePO_4$, together with highly distorted FeO_6 octahedra with C_s symmetry. Principal axes are not identical with unit-cell.

Local Structure

As a short-range characterization, EXAFS provides independent information on the local structures around Fe and Mn[7,9,10]. Of particular interest here is the local structure around Mn^{3+} in the charged state, $(Mn^{3+}_yFe^{3+}_{1-y})PO_4$. In phospho-olivines, the edge-sheared arrangement of the oxygen-octahedra MO_6 (M = Fe, Mn) as well as the localized character of 3d electrons will enhance the selective local lattice distortion around the Jahn-Teller active Mn^{3+} ions. The Fourier transform of $k^3\chi$, $|F(r)|$, shows a radial atomic distribution-like function in real space about the emitting metal ions (Fig. 8). Note that these peaks are at shorter distances than the actual interatomic distance, because the phase shift in the electron-scattering process is not considered.

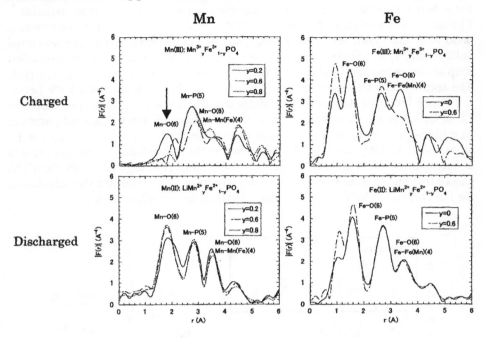

Fig. 8 Radial distribution functions obtained by EXAFS

As shown in Fig. 8, $|F(r)|$ for Fe^{2+} and Fe^{3+} have similar distribution, indicating the local environment is essentially identical. On the other hand, the amplitude of the Mn^{3+}-O peak in $(Mn^{3+}_yFe^{3+}_{1-y})PO_4$ is abnormally low as compared with the Mn^{2+}-O peak in $Li(Mn^{2+}_yFe^{2+}_{1-y})PO_4$, and is significantly reduced as Mn^{3+} content y increases. The peak height is related to the back-scattering of the photoelectrons by the coordinated atoms; the atoms arranged in coherent coordination shells have a larger contribution[9]. When the distances between the absorber atom and the coordination atoms are not uniform, the related peak shows a broadening and an apparent decrease in the peak height due to interference between the real and imaginary part of the spectrum[9]. Consequently, the local lattice distortion around the Jahn-Teller-active Mn^{3+} is much more severe than that expected from the x-ray diffraction data (Fig. 5), and this situation promotes the phase destabilization close to the point $(x, y) = (0, 1)$[7].

Materials for Electrochemical Energy Conversion and Storage

Elastic Tolerance Limit

Under the constant-volume condition of $(Mn^{3+}_yFe^{3+}_{1-y})PO_4$ (Fig. 6, charged state), the elastic energy of the lattice distortion is accumulated as Mn content increases (Fig. 5). In addition, the severe local lattice deformation around Mn^{3+} (Fig. 8) should enhance this tendency. This makes the Mn-rich phase (y>0.7) unstable and unsuitable for battery application. That is to say, there seems an elastic tolerance limit around y=0.7 set by the lattice distortion[5,7]. To confirm this, Mössbauer spectra were measured for $(Mn^{3+}_yFe^{3+}_{1-y})PO_4$ with various manganese contents. Mössbauer measurements were performed because it is very sensitive to minor phases particularly to amorphous phases and/or nanoparticles[4,5]; it is not based on the coherent interaction with the periodic potential of the lattice but involves direct scrutiny of the electronic states around Fe atoms.

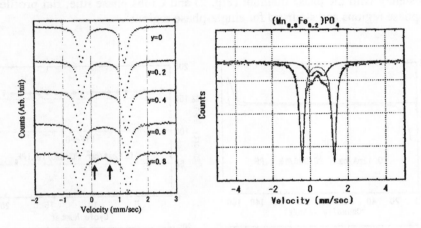

Fig. 9 Mössbauer spectra measured for $(Mn^{3+}_yFe^{3+}_{1-y})PO_4$

The quadrupole splitting (QS), which is the gauge for an asymmetric field around Fe atom and could be confirmed by the first-principle calculation[4,5], increased from 1.53 mm/sec to 1.72 mm/sec as Mn^{3+} content y increase, and note the anomaly observed in a sample with manganese content y=0.8. For samples with smaller manganese content y<0.6, no anomaly was observed and the samples are single-phase olivine. The anomaly for y=0.8 was analyzed as another Fe^{3+} doublet (19.3 %) with isomer shift of 0.42 mm/sec, which means phase separation. This could not be detected by x-ray powder diffraction, in which the amorphous phase and/or nanoparticles contributes to the broad background. Consequently, the elastic tolerance limit is surely in the compositional region between 0.6 and 0.8 [5].

Cathode Performance

Thus, $Li(Mn_{0.6}Fe_{0.4})PO_4$ was chosen[6] as promising composition as 4V cathode and its galvanostatic charge/discharge profiles are shown in Fig. 10, together with those of $LiFePO_4$[4]. The reversible capacity > 160mAh/g, more than 70% of which is obtained under high-rate (C/2) conditions, is almost equal to the theoretical value and cycle very well. The operating voltages of 4.1V for Mn^{3+}/Mn^{2+} and 3.4V for Fe^{3+}/Fe^{2+} were confirmed by ab-initio calculation[4]. Energy density based on the data in Fig. 10 is larger than that of $LiMn_2O_4$ in Wh/l, and surprisingly, larger than that of $LiCoO_2$ in Wh/kg (under the usual cut-off in charge at 4.2V)[4]. Careful sample/electrode preparation as described in the experimental section was essential to achieve excellent cathode performances. The important point is to realize uniform small particles with no residual trivalent phases[4]. Open circuit voltage for samples with various Mn content y showed consistency with the phase diagram (Fig. 2) and Gibbs phase rule; flat profile for two-phase regions and s-curved for single-phase regions[6].

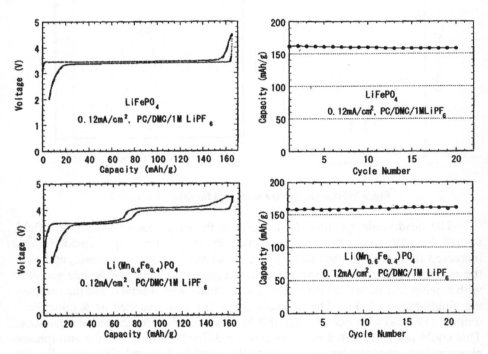

Fig. 10 Charge/discharge profiles and cycle performance at room temperature (23°C) for the optimized electrode composites (90:8:2) of $LiFePO_4$ (top) and $Li(Mn_{0.6}Fe_{0.4})PO_4$ (bottom)

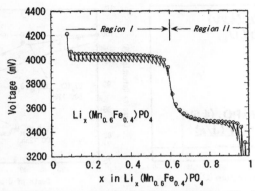

Fig. 11 Open circuit voltage (OCV) of $Li_x(Mn_{0.6}Fe_{0.4})PO_4$

A typical example measured for $Li_x(Mn_{0.6}Fe_{0.4})PO_4$ is shown in Fig. 11. An intermittent discharge mode was used with alternating 20 minutes intervals of continuous discharge at $0.3 mA/cm^2$ and 240 minutes rest intervals, which gave an acceptable equilibrium condition. The overvoltage observed in region I is much larger than that in region II, presumably because, in region I, (1) there is a large resistance for Li^+ to pass through the two-phase interface and (2) the effective mass of the polaronic 3d electrons in the Mn^{3+} phase is large due to the Jahn-Teller effect[6,8]. What should be noticed is, the average voltage $\approx 3.5V$ in region II is considerably higher than that of Li_xFePO_4 ($0<x<1$, $\approx 3.4V$). This is explained by the Fe-O-Mn superexchange interaction, which lowers the Fe^{3+}/Fe^{2+} couple (higher voltage) and raises Mn^{3+}/Mn^{2+} couple (lower voltage) by ca. 0.1V, as previously discussed by Padhi et al[1].

Stability under Unusual Conditions

In the design of consumer batteries, safety issues must be paramount. The large short-circuit current in high-energy lithium batteries raises the local temperature and induces oxygen release from the cathode. This reaction is particularly enhanced in the charged state. The released oxygen ignites the organic electrolyte, which causes reactant gas evolution, and in the worst case, fire and possibly an explosion. In phospho-olivines, all oxygen ions form strong covalent bonds with P^{6+} to form the $(PO_4)^{3-}$ tetrahedral polyanion and consequently difficult to extract. The DSC trace of the fully charged state with electrolyte is consistent with this (Fig. 12). The total heat evolution is only 147 J/g and this is detected over a wide temperature range of 250-360°C[4]. Similar mild reactions were also confirmed for the charged state of Mn/Fe solid-solution system. This guarantees safety against combustion and adds greatly to the attractiveness of the Olivine-type cathode[4].

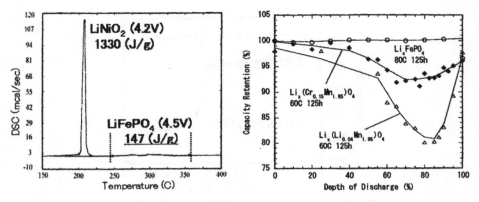

Fig. 12 Left: DSC traces (10°C/min) for the charged LiNiO$_2$ (4.2V) and LiFePO$_4$ (4.5V) with electrolyte. Right: The capacity retention after the high-temperature storage of coin cells at various depth of discharge, for Olivine LiFePO$_4$ (80°C 125h), Spinel Li(Cr$_{0.15}$Mn$_{1.85}$)O$_4$ (60°C 125h), and Spinel Li(Li$_{0.04}$Mn$_{1.96}$)O$_4$ (60°C 125h).

In order to investigate the capacity retention under the long-term, high-temperature storage, the coin cells after 2-cycles of charge/discharge processes were exposed to the high temperature conditions (60°C or 80°C) for 125 hours. Each cell was set to the various depth of discharge (DOD) before storage. After storage, the capacity retention (CR) and self-discharge (SD) were evaluated.

$$CR(\%) = 100C_{AS}/C_{BS}$$
$$SD(\%) = 100[C_{BS}(1-DOD/100) - C_{AS}']/C_{BS}$$

where C_{AS} and C_{BS} are, respectively, the discharge capacity within 4.5-3.0V after/before the storage, and C_{AS}' is the discharge capacity down to 3.0V just after the storage without charging process. As shown in Fig. 12, CR of Li$_x$FePO$_4$ is close to 100% over a wide depth of discharge (DOD in %) even after 125 hours storage at 80°C. These are in contrast to the large capacity loss in the LiMn$_2$O$_4$ - based cathodes even under the lower-temperature (60°C) storage conditions. SD was also negligibly small for Li$_x$FePO$_4$. Similar tolerance was also confirmed for the Li$_x$(Mn$_y$Fe$_{1-y}$)PO$_4$ system.

Strategies

It is readily apparent that the olivine-type LiFePO$_4$ and Li(Mn$_{0.6}$Fe$_{0.4}$)PO$_4$ are very promising materials for lithium battery cathodes. The phase diagram (Fig. 3) leads us to the intuitive idea that Li$_x$MnPO$_4$ (x>0.4) is not entirely inactive in the electrochemical charge-discharge reaction at 4.1V vs Li/Li$^+$ and may be useful as a stable yet compatible 4V cathode for LiCoO$_2$, LiNiO$_2$, and LiMn$_2$O$_4$. However, our electrochemical investigations of $LiMnPO_4$ gave negative results even though the optimized electrode composite was used. Actually, Li_xMnPO_4 showed an open-circuit voltage of at 4.1V, but with much smaller capacity than the theoretical value of 170mAh/g and very large polarization in the galvanostatic charge/discharge mode. In comparative considerations of $LiMnPO_4$ with $LiMn_2O_4$ as a practical 4V cathode, several intrinsic obstacles in $LiMnPO_4$ must be overcome, such as the materials cost, the limited capacity, the much slower reaction kinetics, the need to synthesis in inert gas using divalent source, and a much lower true density (3.4g/cc:$LiMnPO_4$ vs. 4.2g/cc:$LiMn_2O_4$) and the resultant lower energy density. Thus, at the moment, olivine-type $LiMnPO_4$ and Mn-rich $Li(Mn_yFe_{1-y})PO_4$ (y>0.8) are far away from a practical application.

REFERENCES

[1]A. K. Padhi, K. S. Nanjundaswamy, and J. B. Goodenough, "Phospho-olivines as Positive-Electrode Materials for Rechargeable Lithium Batteries," *J. Electrochem. Soc.*, 144[4], 1188-1194 (1997)

[2]A. S. Andersson, B. Kalska, L. Häggström, and J. O. Thomas, "Lithium Extraction/Insertion in $LiFePO_4$: an X-ray diffraction and Mössbauer Spectroscopy Study," Soiid State Ionics, 130, 41-52 (2000)

[3]K. Amine, H. Yasuda, and M. Yamachi, "Olivine $LiCoPO_4$ as 4.8V Electrode Material for Lithium Batteries," Electrochem. Solid State Lett., 3[4], 178-180 (2000)

[4]A. Yamada, S. C. Chung, and K. Hinokuma, "Optimized $LiFePO_4$ for Lithium Battery Cathodes," *J. Electrochem. Soc.*, 148[3], A224-A229 (2001)

[5]A. Yamada and S. C. Chung, "Crystal Chemistry of the Olivine-type $Li(Mn_yFe_{1-y})PO_4$ and $(Mn_yFe_{1-y})PO_4$ as Possible 4V Cathode Materials for Lithium Batteries," *J. Electrochem. Soc.*, 148, to be published (2001)

[6]A. Yamada, Y. Kudo, and K. Y. Liu, "Reaction Mechanism of the Olivine-Type $Li_x(Mn_{0.6}Fe_{0.4})PO_4$," *J. Electrochem. Soc.*, 148[7], in press (2001)

[7]A. Yamada, Y. Kudo, and K. Y. Liu, "Phase Diagram of $Li_x(Mn_yFe_{1-y})PO_4$ (0<x,y<1)," *submitted to J. Electrochem. Soc.*

[8]W. Eventoff, R. Martin, and D. R. Peacor, "The Crystal Structure of Heterosite," American Mineralogist, 57, 45-51 (1972)

[9]I. Nakai and T. Nakagome, "In Situ Transmission X-Ray Absorption Fine Struture Analysis of the Li Deintercalation Process in $Li(Ni_{0.5}Co_{0.5})O_2$," *Electrochem. Solid State Lett.*, 1[6], 259-261 (1998)

[10]H. Yamaguchi, A. Yamada, and H. Uwe, "Jahn-Teller Transition of $LiMn_2O_4$ Studied By X-Ray Absorption Spectroscopy," *Phys. Rev. B*, 58(1), 8-11 (1998)

AMORPHOUS MANGANESE OXIDE CATHODES FOR RECHARGEABLE LITHIUM BATTERIES

D. Im and A. Manthiram
Materials Science and Engineering Program, ETC 9.104
The University of Texas at Austin
Austin, TX 78712

ABSTRACT

Manganese oxides are appealing for lithium-ion cells as they are inexpensive and environmentally benign. With an objective to overcome some of the difficulties encountered with crystalline manganese oxides such as the spinel $LiMn_2O_4$ and layered $LiMnO_2$, amorphous manganese oxide hosts are being investigated. The synthesis of amorphous manganese oxides by reducing lithium permanganate with various organic and inorganic reducing agents such as lithium iodide, methanol, and gaseous hydrogen in aqueous and nonaqueous media and their electrochemical performance are presented. Some of the manganese oxides synthesized by this approach show high capacity with good cyclability and charge efficiency.

INTRODUCTION

Manganese oxides are being considered as a prominent substitute for $LiCoO_2$ cathodes in lithium-ion batteries due to the low material cost and low toxicity compared to those of cobalt. However, the spinel $LiMn_2O_4$ that has been pursued for a long time tends to exhibit capacity fade due to lattice distortions arising from the single electron in the e_g orbital of Mn^{3+} and manganese dissolution from the cathode lattice into the electrolyte.[1,2] On the other hand, the layered $LiMnO_2$ that is isostructural with $LiCoO_2$ tends to transform to spinel-like phases during electrochemical cycling.[3] One way to circumvent the difficulties of lattice distortions encountered in manganese oxides is to develop amorphous manganese oxide hosts. Accordingly, some amorphous manganese oxides have recently been shown to exhibit high capacity with good cyclability.[4-7] Generally, more than one lithium ion per manganese ion could be reversibly extracted/inserted during the charge/discharge process resulting in a significant increase in capacity; capacities of close to 300 mAh/g have been observed. Additionally, the amorphous manganese oxides appear to accommodate the Jahn-Teller distortion more smoothly compared to the crystalline counterparts.[8]

However, it should be noted that amorphous oxides have more probability of having high surface area or active surface than their crystalline counterparts. These surface properties may be closely related to the chemical and electrochemical stability and the manganese dissolution problems. One of the parameters revealing the effect of surface area and activity is the charge efficiency, which is defined as the ratio of discharge capacity to charge capacity in a given cycle. Low charge efficiency will mean either the material itself is not fully reversible or the occurrence of side reactions while the cell is being charged. To be used as a cathode material in lithium-ion cells, high value (> 99 %) of charge efficiency is required because low charge efficiency of cathode is detrimental to the carbon anode that would experience overcharge in every cycle. With an objective to assess these properties, we carefully monitor the charge efficiency of amorphous manganese oxides synthesized with various agents.

Furthermore, previously the amorphous manganese oxides have been synthesized generally by reducing sodium permanganate. Such a process leads to products containing sodium, which in turn may interfere with the lithium insertion/extraction process. To overcome this difficulty, we pursue here the synthesis of amorphous manganese oxides with lithium permanganate.

Owing to the high oxidizing power of permanganates, a variety of organic and inorganic compounds could be employed as reducing agents. In most cases, however, some part of the reagent may remain in the product without being washed out completely. These residues would not only increase the electrode weight and reduce the capacity, but might also cause unwanted side-reactions during the charge/discharge process. With this in mind, we chose reducing agents that are as small in size as possible. Specifically lithium iodide, methanol, and gaseous hydrogen are used as reducing agents and compared.

Selection of solvent for synthesis was based on similar considerations. Among the solvents having high dielectric constant ε and thereby showing good solubility for permanganate salts, acetonitrile and water were used. Acetonitrile was proven to have good chemical and electrochemical stability and known as a labile ligand that can be easily removed even in the case of species having metal-acetonitrile bonds. Water does not have good electrochemical stability, but free water is not expected to remain in the material after firing. Our preliminary tests showed that highly polar solvents such as dimethylformamide ($\varepsilon = 38.3$), dimetylsulfoxide ($\varepsilon = 47.2$), and methyl formamide ($\varepsilon = 189$) reacted with permanganate instantaneously. Propylene carbonate ($\varepsilon = 66.1$) was much better in chemical stability, but some organic residue was observed in the product by IR spectroscopy. We report here the synthesis and electrochemical properties of various amorphous manganese oxides.

EXPERIMENTAL

Lithium permanganate (LiMnO$_4$) supplied by Carus Chemical Company was used as received. Anhydrous lithium iodide was purchased from Alfa Aesar. All other chemicals were reagent grade and used without further purification.

The reduction reactions were carried out by mixing the solutions of lithium permanganate and the reducing agents and stirring the reaction mixtures for a day. When hydrogen gas was used as the reducing agent, a gas mixture consisting of hydrogen and argon in the ratio of 1:9 was bubbled through the lithium permanganate solution. After the reactions, the solid formed was filtered, washed several times with the same solvent used in synthesis, and fired under vacuum for 1 day. Firing temperature was usually 250 °C, but higher temperatures were applied in some experiments as described later.

The amounts of lithium and manganese in the sample were determined with atomic absorption spectroscopy. The iodine content was determined by a redox titration as described elsewhere[4]. The samples were characterized by X-ray powder diffraction recorded in the 2θ range of 10° to 70°.

Electrochemical performances were evaluated with coin cells employing circular cathodes of 2 cm^2 area, metallic lithium anode, 1 M LiClO$_4$ in ethylene carbonate / dimethyl carbonate electrolyte. The cathodes were fabricated by ball milling first a mixture consisting of about 150 mg of the amorphous oxide sample (70 wt%) and conductive carbon black (25 wt%) for 10 minutes to achieve good electrical conductivity and then mixing with polytetrafluoroethylene (PTFE) binder (5 wt%). The electrochemical data were collected with a current density of 0.1 mA/cm^2. To compensate the loss by polarization, when the voltage reached a specific set point, the constant voltage charging mode was applied until the current became less than 0.02 mA/cm^2 or the charged capacity exceeded 105 % of discharged one to prevent extensive overcharge.

RESULTS AND DISCUSSION

Lithium Iodide as a Reducing Agent

Lithium permanganate (0.63 g) solution in acetonitrile (100 mL) was mixed with anhydrous lithium iodide (1.0 g) solution in acetonitrile (100 mL) and stirred. The molar ratio of the reactants was based on the best result obtained with sodium permanganate previously[5]. The empirical formula of the product obtained after firing at 250 °C was found to be Li$_{2.35}$MnO$_y$I$_{0.12}$ (sample A). The X-ray powder diffraction pattern of this sample is shown in Fig. 1a. It shows a few broad reflections, but no reflections corresponding to alkali metal iodate were observed unlike in the case of reduction reactions carried out with sodium permanganate.[6]

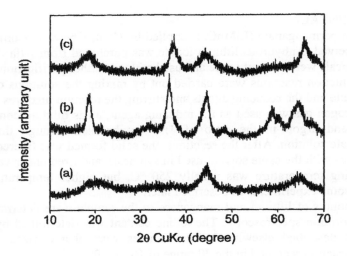

Figure 1.   X-ray powder diffraction patterns of lithium manganese oxides obtained by reducing lithium permanganate with (a) lithium iodide, (b) methanol, and (c) hydrogen followed by firing at 250 °C in vacuum.

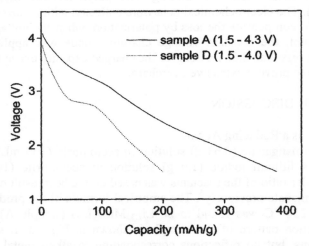

Figure 2. The first cycle discharge profiles of lithium manganese oxides obtained by reducing lithium permanganate with lithium iodide (sample A) and methanol (sample D).

Materials for Electrochemical Energy Conversion and Storage

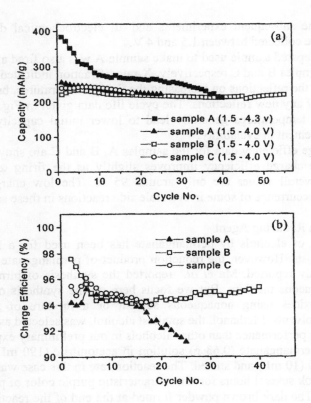

Figure 3. (a) Cycle life and (b) charge efficiency data for samples obtained by reducing lithium permanganate with lithium iodide. Cells were cycled between 1.5 V and 4.0 V except for sample A, for which tests were carried out in two different voltage ranges.

Fig. 2 shows the first cycle discharge profile in the range of 1.5 – 4.3 V. The curve shows a sloping profile with an initial capacity of as high as 385 mAh/g. The high capacity could be due to redox reactions involving manganese and possibly to some extent the iodide as well. The cyclability data of this sample is shown in Fig. 3a. The capacity fades rapidly during the initial 10 cycles and then levels off at around 270 mAh/g. However, the cyclability improves significantly on limiting the cut-off charge voltage to 4 V, but with a decreased initial capacity of 260 mAh/g. Based on this observation, we decided to limit the charge voltage to 4 V in the subsequent experiments and all electrochemical data presented hereafter were collected between 1.5 and 4 V.

The as-prepared sample used to make sample A was also fired at 300 and 350 °C to give samples B and C respectively. X-ray diffraction indicated a growth and narrowing of the reflections on increasing the firing temperatures, but without the appearance of any new reflections. The cycle life data given in Fig. 3 shows that higher firing temperatures generally lead to lower initial capacity, but with a slight improvement in cyclability.

The charge efficiency data for the samples A, B and C are shown in Fig. 3b. Although the charge efficiency improves slightly as the firing temperature is raised, the overall values are only around 95 %. The low charge efficiency suggests the occurrence of some irreversible side reactions in these samples.

Methanol as a Reducing Agent

Oxidation of alcohols by permanganate has been used for a long time by organic chemists. However, the reduction product of permanganate with alcohol has been rarely reported. Ma *et al.*[9] reported the synthesis of birnessites using ethanol in aqueous medium. But we focus here on the synthesis of amorphous manganese oxides using nonaqueous solvent or a mixture of aqueous and nonaqueous solvents. Methanol, the smallest alcohol, was selected as it was found to show better performance than other alcohols in our preliminary experiments.

Lithium permanganate (0.63 g) solution in acetonitrile (190 mL) was mixed with methanol (10 mL) and stirred. The reaction rate in this case was found to be slow and it took several hours for the characteristic purple color of permanganate to disappear. The dark brown powder formed at the end of the reaction was fired at 250 °C in vacuum (sample D). The atomic ratio of lithium to manganese was found to be 1.06. X-ray diffraction pattern and first cycle discharge profile are shown in Figs. 1b and Fig. 2 respectively. Although the peaks are broad, this sample shows better crystallinity than sample A. As a result, a plateau near 3 V emerged in the discharge curve shown in Fig. 2 and the redox process involving this plateau was found to be reversible. The sample shows good cyclability with a capacity of around 200 mAh/g (Fig. 4a). Although the capacity is lower than that found with sample A (Fig. 3a), the charge efficiency that was only 95 % in the initial stage was found to gradually increase and reach 99% after about 30 cycles (Fig. 4b).

The improvement in charge efficiency prompted us to work further on this material. One possibility for the lower charge efficiency during the initial cycles could be the oxidation products of methanol. With an objective to remove the possible polar residues that could be formed during the oxidation reaction of methanol, water was added as a co-solvent during the reaction. With 40 volume % water and 60 volume % acetonitrile in the reaction medium, a manganese oxide

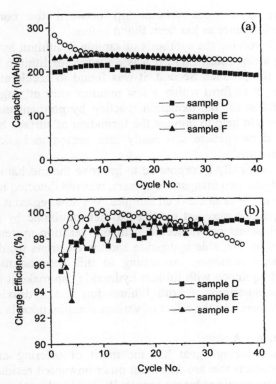

Figure 4. (a) Cycle life and (b) charge efficiency data for samples obtained by reducing lithium permanganate with methanol.

with an Li/Mn ratio of 0.57 was obtained (sample E). Thus the addition of water reduces the lithium content per manganese in the reaction product. Introduction of water also increased the first cycle discharge capacity to 280 mAh/g (Fig. 4a). More importantly, the charge efficiency reached 100 % within a few cycles (Fig. 4b). However, the sample exhibits some capacity decline during the initial cycles (Fig. 4a).

Since the lithium to manganese ratio was low in sample E, we then turned to focus our attention on increasing the lithium content. Towards this objective, we pursued the synthesis in presence of lithium hydroxide (1.05 g) and hydroquinone (0.055 g). The sample obtained after firing at 250 °C (sample F) was found to have an Li/Mn ratio of 1.58. This sample F showed both good cyclability (Fig. 4a) and good charge efficiency of around 99 % after a few cycles. These results suggest that there is a close relationship between the composition and electrochemical performance as has been found before.[5]

In the reaction involving the synthesis of sample F, lithium hydroxide can be considered to play a few roles. First, it was a source for lithium to increase the lithium content in the product. Second, it was found to accelerate the reaction. Normally, solid began to form within a few minutes after mixing. Hydroxide is known to be a catalyst for the oxidation reaction by permanganate in aqueous medium. Third, it might help to prevent the formation of surface hydroxide. Mn-OH with tetravalent manganese can easily lose proton in basic medium and reduce surface hydroxide.[10]

Hydroquinone, originally incorporated to improve the mechanical stability of the material by bridging two manganese centers, was not detected by FT-IR either in the as-prepared sample or in the fired samples. We now regard it as a hydroxide scavenger. Hydroquinone may get oxidized to benzoquinone by giving up two protons and two electrons per molecule.[11] In our reaction system, the protons would be consumed by available hydroxide and the electrons would be provided for the permanganate reduction. According to this reaction mechanism, the combination of hydroquinone with lithium hydroxide can make it possible for us to control the concentrations of both lithium ion and hydroxide during the synthesis. Further research on the effect of various combinations is in progress.

Hydrogen as a Reducing Agent

Hydrogen as a reducing agent has the merit of offering cleaner lithium manganese oxide products that are free from other unwanted residues originating from the organic or inorganic reducing agents. With this advantage in mind, a gas mixture consisting of 10 % H_2 and 90 % Ar was bubbled through an aqueous lithium permanganate solution (1.8 g in 250 mL) for 3 days at 50 °C. The product formed was filtered and fired at 250 °C (sample G). The resultant product was found to have an Li/Mn ratio of only 0.52 that was as low as that in sample E. Electrochemical data collected with this sample are given in Fig. 5. As expected from the low lithium content, the sample was found to exhibit some capacity fade despite a high initial capacity of 300 mAh/g; a capacity loss of 50 mAh/g was observed during the first ten cycles. Charge efficiency (< 94 %) was worse than that found for sample E.

The capacity decline found with this sample could possibly be due to the existence of some hydroxide in the bulk and surface of the sample. Metal hydroxide cannot be easily removed by firing at moderate temperatures like 250 °C. Residual hydroxyl groups might undergo redox reactions as water does, and condensation reaction between two hydroxide groups might block some lithium insertion sites.

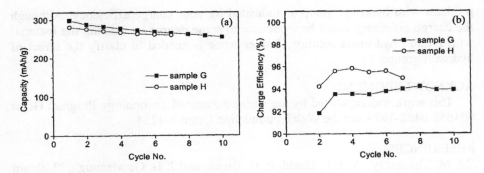

Figure 5. (a) Cycle life and (b) charge efficiency data for samples obtained by reducing lithium permanganate with hydrogen.

In this regard, we tried to convert -OH into -OLi with an ion exchange reaction. The as-prepared sample was dispersed in 0.2 M LiOH aqueous solution and stirred for 1 day at room temperature. Atomic absorption spectroscopic analysis of the fired material (sample H), however, did not show an apparent increase in lithium content. Nevertheless, a comparison of cycle life and charge efficiency data with those of sample G suggests that the hydroxide treatment was somewhat effective in improving the performance. Further ion exchange experiments under more severe conditions are being pursued to have further understanding of these materials.

CONCLUSIONS

Amorphous manganese oxides have been synthesized by reducing lithium permanganate with lithium iodide, methanol, and hydrogen. Materials obtained with lithium iodide generally showed high capacity with good cyclability, but with poor charge efficiency (~95 %). Using methanol as a reducing agent, we were able to get much better charge efficiency although at the sacrifice of the initial capacity and cyclability. However, with careful selection of solvents and additives for the synthesis, both the capacity and cyclability could be improved. Aqueous-nonaqueous mixed solvent systems appear to efficiently wash out the unnecessary lithium salts and organic residues in the reduction product. Additionally, the combination of LiOH and hydroquinone made it possible to control the composition of the products. An amorphous manganese oxide showed quite high initial capacity (~230 mAh/g) and excellent reversibility with charge efficiency of over 99 % when the cells are cycled between 1.5 V and 4.0 V at 0.1 mA/cm^2. The reaction of permanganate with hydrogen appears to produce products with hydroxyl group as indicated by poor charge efficiency. Although the charge efficiency could be overcome to some degree by treating the material with lithium hydroxide solution, further work is needed to clarify the effect of hydroxyl groups.

ACKNOWLEDGMENT

This work was supported by the Texas Advanced Technology Program Grant 003658-0488-1999 and the Welch Foundation Grant F-1254.

REFERENCES

[1] M. M. Thackeray, W. I. F. David, P. G. Bruce, and J. B. Goodenough, "Lithium insertion into manganese spinels," *Mater. Res. Bull.* **18**, 461 (1983).

[2] R. J. Gummow, A. de Kock, and M. M. Thackeray, "Improved capacity retention in rechargeable 4 V lithium/lithium manganese (spinel) cells," *Solid State Ionics* **69**, 59 (1994).

[3] M. M. Thackeray, "Structural considerations of layered and spinel lithiated oxides for lithium ion batteries," *J. Electrochem. Soc.* **142**, 2558 (1995).

[4] J. Kim and A. Manthiram, "A manganese oxyiodide cathode for rechargeable lithium batteries," *Nature* **390**, 265 (1997).

[5] J. Kim and A. Manthiram, "Amorphous manganese oxyiodides exhibiting high lithium intercalation capacity at higher current density," *Electrochem. Solid-State Lett.* **2**, 55 (1999).

[6] A. Manthiram, J. Kim, and S. Choi, "Solution-based synthesis of manganese oxide cathodes for lithium batteries," *Mater. Res. Soc. Symp. Proc.* **575**, 9 (2000).

[7] J. J. Xu, A. J. Kinser, B. B. Owens, and W. H. Smyrl, "Amorphous manganese dioxide: A high capacity lithium intercalation host", *Electrochem. Solid-State Lett.* **1**, 1 (1998).

[8] C. R. Horne, U. Bergmann, J. Kim, K. A. Streibel, A. Manthiram, S. F. Cramer and E. J. Cairns, "Structural Investigations of $Li_{1.5+x}Na_{0.5}MnO_{2.85}I_{0.12}$ Electrodes by Mn X-ray Absorption Near Edge Spectroscopy," *J. Electrochem. Soc.* **147**, 395 (2000).

[9] Y. Ma, J. Luo, and S. L. Suib, "Synthesis of birnessites using alcohols as reducing reagents: Effects of synthesis parameters on the formation of birnessites," *Chem. Mater.* **11**, 1972 (1999).

[10] R. Chitrakar, H. Kanoh, Y. Yoshitaka, and K. Ooi, "Synthesis of spinel-type lithium antimony manganese oxides and their Li^+ extraction/ion insertion reactions," *J. Mater. Chem.* **10**, 2325 (2000).

[11] A. J. Bard and L. R. Faulkner, "Electrochemical Methods"; pp.699, John Wiley and Sons, New York, 1980.

SYNTHESIS AND ELECTROCHEMICAL PROPERTIES OF SPINEL LiCo$_2$O$_4$ CATHODES

S. Choi and A. Manthiram
Materials Science and Engineering Program, ETC 9.104
The University of Texas at Austin
Austin, TX 78712

ABSTRACT

Spinel LiCo$_2$O$_4$ oxide has been synthesized by chemically extracting 50% of lithium ions with the oxidizer Na$_2$S$_2$O$_8$ from the LT-LiCoO$_2$ sample that was synthesized at 400 °C. The samples are characterized by X-ray powder diffraction, atomic absorption spectroscopy for lithium contents, and a redox iodometric titration for oxygen contents. Rietveld refinement of the X-ray diffraction data of the as-prepared and 200 °C heated Li$_{0.5}$CoO$_2$ samples indicate the formation of the normal cubic spinel structure with a cation distribution of (Li)$_{8a}$[Co$_2$]$_{16d}$O$_4$. The open-circuit voltages of the as-prepared and 200 °C heated samples have been collected for various values of x in Li$_{1\pm x}$Co$_2$O$_4$. While the extraction of lithium from the 8a sites occurs at around 3.9 V, the insertion of lithium into the 16c sites occurs at around 3.6 V. The 0.3 V difference observed between the two processes is much smaller than that predicted from theoretical calculations (1.3 V) in the literature and observed for the LiMn$_2$O$_4$ system (1 V).

INTRODUCTION

The exponential growth in portable electronic devices has created an ever-increasing demand for high energy density batteries. Lithium-ion batteries have become appealing to meet this demand due to their higher energy density originating from a higher cell voltage (4 V) compared to other rechargeable systems such as nickel-cadmium and nickel-metal hydride batteries (< 2 V). For lithium-ion cells, the layered LiMO$_2$ (M = Co and Ni) and the spinel LiMn$_2$O$_4$ cathodes offering 4 V have become attractive. In the quest for designing new electrode materials with high energy density, a fundamental understanding of the factors that control the cell voltage has drawn much attention in recent years both from scientific and technological points of view.[1-6]

The open circuit voltage (OCV) profile of a cell is determined by the electrochemical potential difference of the lithium ion in the cathode and anode. The electrochemical potential includes contributions of various energy terms such as the energies involved in electron transfer and the site energies involved in lithium transfer, which is related to the crystal structure, the coordination geometry of the site into/from which the lithium ions are inserted/extracted, and activation energies. For example, the layered $LiCoO_2$ and $LiNiO_2$ exhibit about 4 V versus metallic lithium anode, while the $LiMn_2O_4$ spinel exhibits two plateaus, one around 3 V and the other around 4 V. The 3 V and 4 V regions of $LiMn_2O_4$ correspond to the lithium insertion/extraction into/from the 16c octahedral and 8a tetrahedral sites, respectively, of the spinel lattice. With the same $Mn^{3+/4+}$ couple, the 1 V difference found in the case of spinel $LiMn_2O_4$ has been explained by the site energy difference between the 8a and 16c sites.[7] On the other hand, with a single type of site (octahedral) involved for lithium insertion/extraction, the layered $LiCoO_2$ and $LiNiO_2$ exhibit continuous voltage profiles around 4 V for the $Co^{3+/4+}$ and $Ni^{3+/4+}$ couples.

Unfortunately, the discharge/charge profiles of the spinel framework have been investigated only with the $LiMn_2O_4$ spinel and cation-substituted manganese spinel oxides $LiMn_{2-y}M_yO_4$ (M = Cr, Fe, Co, Ni, Cu and Li) with y ≤ 1;[8,9] no manganese-free LiM_2O_4 (M = transition metal other than Mn) spinel has been experimentally investigated with respect to obtaining the voltage profiles. The cation-substituted $LiMn_{2-y}M_yO_4$ (M = Cr, Fe, Co, Ni and Cu) spinel oxides generally show two plateaus, one at around 4 V and the other at around 5 V, both corresponding to the extraction/insertion of lithium ions from/into the 8a tetrahedral sites. However, a satisfactory understanding of the origin of the 5 V capacity for the above series of spinel oxides has encountered some difficulties. Ohzuku et al[8] attribute the 5 V redox potential to the compact crystal fields imposed by the spinel structure of cubic close-packed oxide ions bonded to tetravalent manganese ions. Kawai et al[9] attribute the 5 V redox potential of, for example, $LiMnCoO_4$ to the increase in site energy (tetrahedral site lithium) compared to that in layered $LiCoO_2$ (octahedral site lithium) or changes in the electronic structure of octahedral site cobalt ions induced by the presence of manganese ions. Thus the origin of the 5 V capacity of the $LiMn_{2-y}M_yO_4$ spinel oxides is still being debated.

In this regard, an investigation of the voltage profiles of manganese-free spinel oxides such as $LiCo_2O_4$ and $LiNi_2O_4$ may be able to give new insight in understanding the origin of the 5 V capacity. Unfortunately, the difficulties in synthesizing such manganese-free LiM_2O_4 (M = transition metal) spinel oxides by conventional procedures have largely prevented such investigations. However, the voltage profiles of spinel $LiCo_2O_4$ have been predicted from theoretical calculations in the literature. Theory has predicted a two-step voltage profile for

LiCo$_2$O$_4$: one around 3 V and the other at 4.3 – 5.3 V corresponding to the lithium extraction/insertion from/into the 16c octahedral and 8a tetrahedral sites respectively.[5,6]

The normal spinel LiCo$_2$O$_4$ is, in fact, known to be formed on extracting lithium from the low temperature form of LiCoO$_2$ (designated as LT-LiCoO$_2$), which is synthesized at 400 °C in air.[10-13] However, no experimental voltage profiles are available in the literature for LiCo$_2$O$_4$. We present in this paper the synthesis of the normal spinel LiCo$_2$O$_4$ by chemically extracting 50% of lithium with an oxidizing agent Na$_2$S$_2$O$_8$ from the LT-LiCoO$_2$. Additionally, the structural stability of the LiCo$_2$O$_4$ spinel on heating and its voltage profiles on extracting/inserting lithium is presented.

EXPERIMENTAL

The LT-LiCoO$_2$ sample was synthesized by firing required amounts of Li$_2$CO$_3$ and Co$_3$O$_4$ at 400 °C in air for 1 week to ensure the completion of the reaction. Chemical extraction of lithium from the LT-LiCoO$_2$ sample to obtain Li$_{0.5}$CoO$_2$ was carried out by stirring the LiCoO$_2$ powders for two days with an aqueous solution consisting of optimized amounts of the oxidizing agent Na$_2$S$_2$O$_8$. During this process, the following reaction occurred:

$$2 \text{ LiCoO}_2 + x \text{ Na}_2\text{S}_2\text{O}_8 \rightarrow 2 \text{ Li}_{1-x}\text{CoO}_2 + x \text{ Na}_2\text{SO}_4 + x \text{ Li}_2\text{SO}_4 \qquad (1)$$

The product formed was filtered, washed repeatedly first with water and finally with acetone, and air-dried. The samples were characterized by X-ray powder diffraction. Structural information and lattice parameters were obtained by Rietveld refinement of the X-ray data with the DBWS-9411 PC program.[14] Lithium contents were determined by atomic adsorption spectroscopy. Oxygen contents were determined by a redox (iodometric) titration.[15] Electrochemical properties were evaluated with coin cells using circular cathodes of 0.64 cm^2 area, metallic Li anodes, and LiPF$_6$ in diethyl carbonate/ethylene carbonate electrolyte. The cathodes were fabricated by hand mixing the fired samples with 25 wt % fine carbon for about 30 minutes and blending the hand mixed composite with 5 wt % polytetrafluoroethylene. The discharge/charge curves were recorded with a current density of 0.36 mA/cm^2. Open-circuit voltages at each discharge/charge step were obtained after allowing the cells to rest for 3 days.

RESULTS AND DISCUSSION
Synthesis and the characterization

Fig. 1 shows the X-ray diffraction patterns of the LT-LiCoO$_2$ sample before and after extracting lithium. LT-LiCoO$_2$ is known to adopt a structure that is different from the structure of the conventional high temperature form that has a

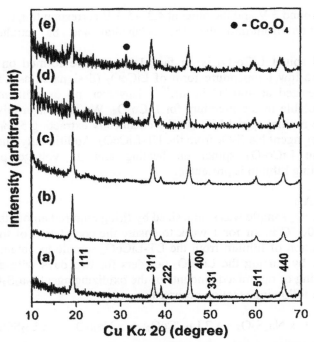

Fig. 1 X-ray diffraction patterns of $LiCoO_2$ synthesized at 400 °C and $Li_{0.5}CoO_2$ obtained from it: (a) $LiCoO_2$, (b) as-prepared $Li_{0.5}CoO_2$ (spinel), (c) $Li_{0.5}CoO_2$ after heating at 200 °C (spinel), (d) $Li_{0.5}CoO_2$ after heating at 300 °C, and (e) $Li_{0.5}CoO_2$ after heating at 400 °C.

layered rhombohedral structure. Thackeray's group has investigated the structure of LT-LiCoO$_2$ using X-ray and neutron diffractions and concluded that the LT-form has lithiated spinel-like structure.[10-12] Recently, they have also shown that the LT-form consists of two phases: a dominant lithiated spinel phase $\{Li\}_{16c}[Co_2]_{16d}O_4$ and a minor layered phase. They showed further with transmission electron microscopy and electron diffraction studies that an extraction of lithium with acid gives a single normal spinel phase $(Li)_{8a}[Co_2]_{16d}O_4$.[13] The pattern in Fig. 1a confirms the formation of LT-LiCoO$_2$.

Figs. 1b and 1c show the X-ray patterns of $Li_{0.5}CoO_2$, which was obtained by extracting lithium from LT-LiCoO$_2$, before and after heating at 200 °C. Fig. 2 shows the Rietveld refinement results of the as-prepared and 200 °C heated $Li_{0.5}CoO_2$. The refinement confirms the formation of the normal spinel structure

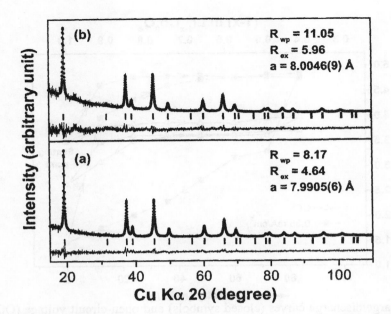

Fig. 2 Rietveld refinement results of $Li_{0.5}CoO_2$: (a) as-prepared $Li_{0.5}CoO_2$ obtained from 400 °C LT-LiCoO_2, (b) sample (a) after heating at 200 °C. The observed and calculated X-ray profiles, peak positions, and the difference between the observed and calculated profiles are shown.

in both cases as indicated by a good matching between the observed and calculated patterns. Atomic absorption spectroscopy results show a lithium to cobalt ratio of 1.04 : 2.0 confirming the spinel composition of $LiCo_2O_4$. The iodometric redox titration indicates an oxygen content of 3.94 for the as-prepared sample with a small oxygen deficiency of $\delta = 0.06$ in $LiCo_2O_{4-\delta}$. The 200 °C heated sample has an oxygen content of 3.89 with $\delta = 0.11$. In spite of their slight anion deficiency, both the samples maintain the spinel framework as revealed by the Rietveld refinements in Fig. 2. However, heating above 200 °C results in a disproportionation of the spinel phase into a mixture of $LiCoO_2$ and Co_3O_4 in accordance with the following reaction as indicated by the X-ray data in Fig 1(d) and 1(e):

$$6 Li_{0.5}CoO_2 \rightarrow 3 LiCoO_2 + Co_3O_4 + O_2 \qquad (2)$$

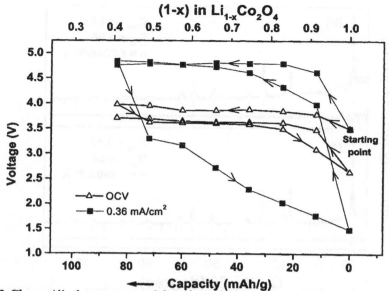

Fig. 3 Charge/discharge curves (closed symbols) and open-circuit voltage (OCV) curves (open symbols) of 200 °C heated LiCo$_2$O$_4$ spinel. The data correspond to the extraction/insertion of lithium ions from/into the 8a tetrahedral sites.

The oxygen loss from LiCo$_2$O$_{4-\delta}$ on heating was also confirmed by thermogravimetric analysis (TGA). Thus the LiCo$_2$O$_{4-\delta}$ spinel oxides are metastable and can be accessed only by low temperature soft chemistry techniques.

Electrochemical study

The two normal spinel samples discussed above (as-prepared and 200 °C heated) were subjected to the electrochemical test with coin-type cell as described in the experimental section. Both samples are found to exhibit large polarization loss during the charge/discharge process as shown in Figs. 3 - 5. As a result, we decided to collect open-circuit voltages (OCV) at various depths of discharge/charge corresponding to the extraction/insertion of lithium ions from/into the 8a tetrahedral and 16c octahedral sites. As seen in Fig. 3, the OCV for the process of lithium extraction/insertion from/into 8a tetrahedral sites of the 200 °C heated LiCo$_2$O$_4$ sample to give Li$_{1-x}$CoO$_2$ is around 3.9 – 4.0 V for the range of $0.4 \leq (1-x) \leq 1$. This experimentally observed OCV range is quite lower than the theoretically calculated values of 4.3 – 5.3 V.[5,6] However, after the first charge, the OCV for the subsequent discharge/charge processes is slightly lower

Materials for Electrochemical Energy Conversion and Storage

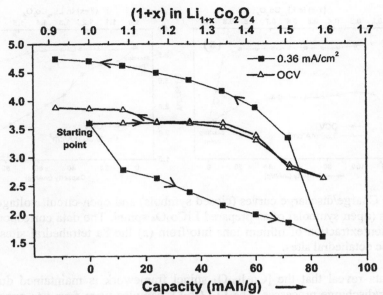

Fig. 4 Charge/discharge curves (closed symbols) and open-circuit voltage (OCV) curves (open symbols) of 200 °C heated LiCo$_2$O$_4$ spinel. The data correspond to the insertion/extraction of lithium ions into/from the 16c octahedral sites.

as seen in Fig. 3. This difference may indicate some irreversible changes during the first charge. One possibility is a loss of some oxygen from the lattice during the first charge. For example, the layered Li$_{1-x}$CoO$_2$ cathodes have been shown to lose oxygen from the lattice at low lithium contents (1-x) < 0.5.[16,17] Nevertheless, a good matching between the OCV profiles of the subsequent cycles (first discharge and second charge) suggests good reversibility for the lithium extraction/insertion process. Similar OCV results are also obtained for the as-prepared LiCo$_2$O$_4$ as shown in Fig. 5(a).

Fig. 4 shows the discharge/charge curves and the OCV curves corresponding to the insertion/extraction of lithium ions into/from the 16c octahedral sites of the 200 °C LiCo$_2$O$_4$ to give Li$_{1+x}$Co$_2$O$_4$ for the range of $0.9 \leq (1+x) \leq 1.6$. As seen from the OCV data in Fig. 4, this process occurs at around 3.6 V and it is reversible. Similar voltage profile is observed for the as-prepared LiCo$_2$O$_4$ spinel also as seen in Fig. 5(b).

In order to further check the reversibility of the lithium insertion/extraction processes involving 8a and 16c sites, we have also examined the cathodes by X-ray diffraction after the charge/discharge processes. Rietveld refinement of the X-

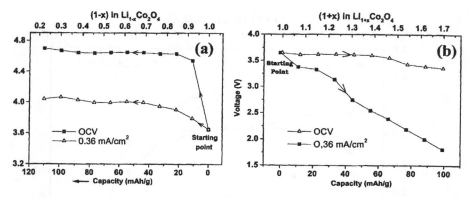

Fig. 5 Charge/discharge curves (closed symbols) and open-circuit voltage (OCV) curves (open symbols) of as-prepared LiCo$_2$O$_4$ spinel. The data correspond to the insertion/extraction of lithium ions into/from (a) the 8a tetrahedral sites and (b) the 16c octahedral sites.

ray data reveal that the [Co$_2$]$_{16d}$O$_4$ spinel framework is maintained during the charge/discharge processes. The Li$_{1-x}$Co$_2$O$_4$ samples were found to experience a decrease in lattice parameter due to the oxidation of Co^{3+} to Co^{4+} and the Li$_{1+x}$Co$_2$O$_4$ samples were found to experience an increase in lattice parameter due to a reduction of Co^{4+} to Co^{3+}.

The experimental electrochemical data of LiCo$_2$O$_4$ spinel oxides reveal that the lithium extraction/insertion from/into the 8a tetrahedral and 16c octahedral sites occur, respectively, at 3.9 – 4.0 V and 3.6 V. These values are quite different from the theoretically predicted values of 4.3 – 5.3 V for the 8a sites and 3 V for the 16c sites. The lithium extraction/insertion processes into the 8a and 16c sites are thus separated by a much smaller voltage difference (0.3 V) than that predicted by theoretical calculations for LiCo$_2$O$_4$ (1.3 V) and observed experimentally for the LiMn$_2$O$_4$ spinel system (1 V). However, the possible loss of some oxygen during the first charge may have some effect on the experimental voltage profile. It is possible that the Jahn-Teller distortion occurring in the Li$_{1+x}$Mn$_2$O$_4$ system may play a role in the observed 1 V drop. Additionally, manganese-containing octahedra may play a role in imparting the 5 V capacity for LiMn$_{2-y}$M$_y$O$_4$ (M = Fe, Co, Ni, Cu) spinel oxides. Clearly, investigation of other LiM$_2$O$_4$ spinel systems is needed to develop a better understanding of the voltage profiles of spinel oxides.

CONCLUSIONS

$LiCo_2O_4$ spinel oxide has been synthesized by chemically extracting lithium with an oxidizing agent $Na_2S_2O_8$ from LT-$LiCoO_2$ that was synthesized at 400 °C. The material was found to have the normal spinel structure with a cation distribution of $(Li)_{8a}[Co_2]_{16d}O_4$. The $LiCo_2O_4$ spinel, however, disproportionates on heating at $T > 200$ °C to give the thermodynamically more stable phases $LiCoO_2$ and Co_3O_4. The voltage profiles corresponding to the lithium extraction/insertion processes from/into the 8a tetrahedral and 16c octahedral sites of the $Li_{1\pm x}Co_2O_4$ spinel have been experimentally investigated. The experimentally observed voltages for these processes are different from those predicted from theoretical calculations in literature. Experimentally, a much smaller voltage separation (0.3 V) is observed between the two processes compared to the theoretically predicted value of 1.3 V.

ACKNOWLEDGMENT

This work was supported by the Welch Foundation Grant F-1254, Texas Advanced Technology Program Grant 003658-0488-1999, and Center for Space Power at the Texas A&M University (a NASA Commercial Space Center).

REFERENCES

[1]A. Manthiram and J. B. Goodenough, "Lithium Insertion into $Fe_2(SO_4)_3$-Type Frameworks," *J. Power Sources,* **26**, 403 (1989).

[2]J. B. Goodenough, "Design Considerations," *Solid State Ionics,* **69**, 184 (1994).

[3]M. K. Aydino and G. Ceder, "First-Principles Prediction of Insertion Potentials in Li-Mn Oxides for Secondary Li Batteries," *J. Electrochem. Soc.,* **144**, 3832 (1997).

[4]A. Manthiram and J. Kim, "Low Temperature Synthesis of Insertion Oxides for Lithium Batteries," *Chem. Mater.,* **10**, 2895 (1998).

[5]C. Wolverton and A. Zunger, "Prediction of Li intercalation and Battery Voltage in Layered vs. Cubic Li_xCoO_2," *J. Electrochem. Soc.,* **145**, 2424 (1998).

[6]A. Van der Ven and G. Ceder, "Electrochemical Properties of Spinel Li_xCoO_2: A First-principles Investigation," *Phys. Rev.,* **B59**, 742 (1999).

[7]M. M. Thackeray, "Manganese Oxides for Lithium Batteries," *Prog. Solid State Chem.,* **25**, 1 (1997).

[8]T. Ohzuku, S. Takeda, and M. Iwanaga, "Solid-state Redox Potentials for $Li[Me_{1/2}Mn_{3/2}]O_4$ (Me: 3d Transition Metal) having Spinel Framework Structures: a Series of 5 Volt Materials for Advanced Lithium-ion Batteries," *J. Power Sources,* **81-82**, 90 (1999).

[9]H. Kawai, M. Nagata, H. Tukamoto, and A.R. West, "High-voltage Lithium Cathode Materials," *J. Power Sources,* **81-82**, 67 (1999).

[10]R.J. Gummow, M.M. Thackeray, W.I.F. David, and S. Hull, "Structure and Electrochemistry of Lithium Cobalt Oxide Synthesized at 400 °C," *Mat. Res. Bull.*, **27**, 327 (1992)

[11]R.J. Gummow, D.C. liles, and M.M. Thackeray, "Spinel Versus Layered Structures for Lithium Cobalt Oxide Synthesized at 400 °C," *Mat. Res. Bull.*, **28**, 235 (1993)

[12]R.J. Gummow, D.C. liles, and M.M. Thackeray, "A Reinvestigation of the Structures of Lithium-Cobalt-Oxides with Neutron-Diffraction Data," *Mat. Res. Bull.*, **28**, 1177 (1993)

[13]Y. Shao-Horn, S.A. Hackney, C.S. Johnson, A.J. Kahaian, and M.M. Thackeray, "Structural Features of Low-Temperature $LiCoO_2$ and Acid-Delithiated Products," *J. Solid State Chem.*, **140**, 116 (1998)

[14]R. A. Young, A. Shakthivel, T. S Moss, and C. O. Paiva Santos, DBWS-9411 Program for Rietveld Refinement," *J. Appl. Cryst.*, **28**, 366 (1995).

[15]A. Manthiram, S. Swinnea, Z. T. Sui, H. Steinfink, and J. B. Goodenough, "The Influence of Oxygen Variation on the Crystal Structure and Phase Composition of the Superconductor $YBa_2Cu_3O_{7-x}$," *J. Amer. Chem. Soc.*, **109**, 6667 (1987).

[16]R. V. Chebiam, A. M. Kannan, F. Prado, and A. Manthiram,"Comaprision of the Chemical Stability of the High Energy Density Cathodes of Lithium-ion Batteries," *Electrochem. Commun.* (in press).

[17]R. V. Chebiam, F. Prado, and A. Manthiram, "Soft Chemistry Synthesis and Characterization of Layered $Li_{1-x}Ni_{1-y}Co_yO_{2-\delta}$," *Chem. Mater.* (in press).

DESIGNING STRUCTURALLY STABLE LAYERED OXIDE CATHODES FOR LITHIUM-ION BATTERIES

S. Choi and A. Manthiram
Materials Science and Engineering Program, ETC 9.104
The University of Texas at Austin
Austin, TX 78712

ABSTRACT

With an aim to design structurally stable layered cathode hosts for lithium-ion batteries, the ease of transformation of $Li_{0.5}MO_2$ (M = Mn, Co, Ni, $Mn_{1-y}Cr_y$, $Mn_{1-y}(Cr,Al,Mg)_y$, $Ni_{1-y}Co_y$ and $Ni_{1-y}Mn_y$) compositions to cubic spinel-like phases has been investigated systematically. The structural stability is assessed by chemically extracting 50% of lithium ions from the layered $LiMO_2$ oxides with the oxidizing agent sodium perdisulfate ($Na_2S_2O_8$) to obtain $Li_{0.5}MO_2$ followed by heating at various temperatures. It is found that $Li_{0.5}MnO_2$ transform easily to spinel-like phases at ambient temperatures during the process of lithium extraction while the $Li_{0.5}NiO_2$ composition needs some heat (T $\approx$ 200 °C) for the transformation and the $Li_{0.5}CoO_2$ composition does not transform even at T $\approx$ 200 °C. The differences among the three systems are explained on the basis of crystal field stabilization energies. However, substitution of other cations such as Cr and (Cr,Al,Mg) for Mn is found to be effective in suppressing the transformation.

INTRODUCTION

Lithium-ion batteries have become attractive for popular portable electronic devices such as cellular phones and laptop computers due to their higher energy density compared to other systems. The commercial lithium-ion cells currently use the layered $LiCoO_2$ as the cathode[1] and carbon as the anode. Unfortunately, cobalt is expensive and toxic. Moreover, only 0.5 lithium per Co could be reversibly extracted/inserted during the charge/discharge process, which limits its practical capacity to approximately 140 mAh/g. These drawbacks have stimulated immense interest to develop alternate cathode materials for lithium-ion cells. In this regard, spinel $LiMn_2O_4$,[2] layered $LiMnO_2$,[3,4] orthorhombic $LiMnO_2$,[5] and layered $LiNiO_2$[6] have become appealing as Mn is inexpensive and environmentally friendly and Ni is less expensive and less toxic than Co.

However, the $LiMn_2O_4$ spinel is confronted with capacity fade originating from manganese dissolution from the lattice into the electrolyte and Jahn-Teller distortion[7,8] and both the layered and orthorhombic $LiMnO_2$ tend to transform to spinel-like phases during electrochemical cycling.[3-5] On the other hand, $LiNiO_2$ suffers from safety concerns and formation of spinel-like phases under mild heat due to a migration of nickel ions to the lithium planes.[9]

The transformation of layered phases into spinel-like phases has also been predicted by theoretical calculations based on first principles. The calculations show that the spinel structure is energetically more stable than the layer structure for a variety of $Li_{0.5}MO_2$ (M = Ti, Cr, Mn, Fe, Co, and Ni) compositions.[10] The layered to spinel-like phase transition becomes plausible as both the structures have a cubic close-packed oxygen array. With the same type of oxide-ion packing, the transformation of the rhombohedral layered $(Li_{0.5})_{3a}(M)_{3b}O_2$ into the cubic spinel phase $(Li)_{8a}[M_2]_{16d}O_4$ requires a migration of 25% of the transition metal ions M^{n+} from the original M planes (3b sites) into the lithium planes (3a sites) and a displacement of the lithium ions from the 3a octahedral sites into the neighboring tetrahedral sites. The migration of the transition metal ions M^{n+} from the M plane to the lithium plane can occur via the available neighboring tetrahedral sites (designated as T_1 and T_2) as described in Fig. 1.

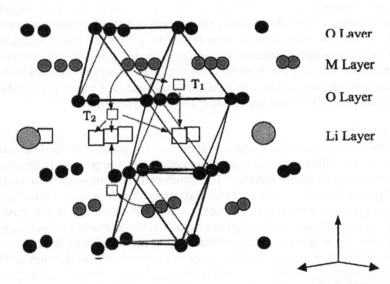

Fig. 1 Illustration of the paths for the migration of the transition metal ion from the octahedral sites in the M planes to the octahedral sites in the lithium planes via the neighboring tetrahedral sites (T_1 and T_2).

However, a systematic experimental study of the formation of spinel phases from the layered phases for various $Li_{0.5}MO_2$ (M = transition metal) compositions is lacking in the literature excepting for $Li_{0.5}NiO_2$ which has been reported to transform to spinel under mild heat.[11] We present in this paper, a systematic investigation and comparison of the layered to spinel-like phase transitions in chemically synthesized $Li_{0.5}MO_2$ (M = Mn, Co, Ni, $Mn_{1-y}Cr_y$, $Mn_{1-y}(Cr,Al,Mg)_y$, $Ni_{1-y}Co_y$ and $Ni_{1-y}Mn_y$). The ease of transformation is assessed by investigating the crystal chemistry of the $Li_{0.5}MO_2$ phases before and after heating at various temperatures; the $Li_{0.5}MO_2$ phases are synthesized by chemically extracting lithium from the corresponding $LiMO_2$ at ambient temperature with an oxidizing agent.

EXPERIMENTAL

$LiCoO_2$ was synthesized by solid state reactions between required amounts of Li_2CO_3 and Co_3O_4 at 800 °C. $LiNi_{1-y}Co_yO_2$ samples with $0 \le y \le 1$ were synthesized by sol-gel procedure described below. Required amounts of Li_2CO_3, cobalt acetate and nickel acetate were dissolved in acetic acid and the mixture was refluxed to obtain a clear solution. After refluxing for one hour, small amounts of 30 % H_2O_2 and water were added and the mixture was refluxed for another hour. The resulting sol was then allowed to dry slowly on a hot plate to give a gel. The gel so obtained was then decomposed at 400 °C and fired at 800 °C under flowing oxygen for 24 hours. $LiNi_{1-y}Mn_yO_2$ samples with $0 \le y \le 0.5$ were synthesized by firing a coprecipitated hydroxides of nickel and manganese with lithium hydroxide at 800 °C under flowing oxygen. The coprecipitation of the nickel and manganese hydroxides were achieved by adding KOH to the corresponding acetate solution followed by filtering, washing and drying of the product. Orthorhombic $LiMnO_2$ was synthesized by solid state reaction between Li_2CO_3 and Mn_2O_3 at 1000 °C under flowing nitrogen. Layered $LiMnO_2$ (monoclinic) was obtained by ion-exchanging α-$NaMnO_2$ with LiBr in hexanol at 150 °C. Monoclinic $LiMn_{1-y}M_yO_2$ (M = Cr or (Cr,Al)) samples were synthesized by solid state reactions of required amounts of LiOH, Mn_2O_3, Cr_2O_3, Al_2O_3, and $Mg(OH)_2$ at 1050 °C under flowing nitrogen with one intermittent grinding.

Chemical extraction of lithium from the various $LiMO_2$ phases to obtain $Li_{0.5}MO_2$ was accomplished by stirring for two days the $LiMO_2$ powders with an aqueous solution containing appropriate quantities of the oxidizer, sodium perdisulfate ($Na_2S_2O_8$). The following chemical reaction occurs during this process:

$$2\,LiMO_2 + 0.5\,Na_2S_2O_8 \rightarrow 2\,Li_{0.5}MO_2 + 0.5\,Na_2SO_4 + 0.5\,Li_2SO_4 \quad [1]$$

After the chemical extraction reaction, the product $Li_{0.5}MO_2$ formed was washed with deionized water several times to remove the soluble Na_2SO_4 and Li_2SO_4 and any unreacted $Na_2S_2O_8$. The $Li_{0.5}MO_2$ samples thus obtained were then heated in air for 3 days at various temperatures to assess the ease of formation of spinel-like phases. Lithium contents in the product were determined by atomic adsorption spectroscopy. Crystal-chemical characterizations were carried out by fitting the X-ray diffraction data with the Rietveld program.

RESULTS AND DISCUSSION

Layered to Spinel-like Phase Transformation in $Li_{0.5}MO_2$ (M = Mn, Co, Ni, $Ni_{1-y}Co_y$, and $Ni_{1-y}Mn_y$)

The X-ray diffraction pattern of chemically prepared $Li_{0.5}NiO_2$ is compared with that of $LiNiO_2$ in Fig. 2. It also gives the evolution of phases on heating $Li_{0.5}NiO_2$ at various temperatures. Although the as-prepared $Li_{0.5}NiO_2$ (Fig. 2b) maintains the rhombohedral layer structure of the initial $LiNiO_2$, it transforms to a cubic phase on heating at 200 °C as indicated by the merger of the (018) and (110) reflections centered around $2\theta = 65°$ into a single reflection (Fig. 2d). The cubic phase begins to disproportionate with a loss of oxygen to give $LiNiO_2$ and NiO on heating above 200 °C (Fig. 2e). Although the formation of a normal spinel phase $(Li)_{8a}[Ni_2]_{16d}O_4$ has been claimed in the literature before,[11] our Rietveld analysis with the space group Fd3m shows that the cubic phases obtained at 200 °C

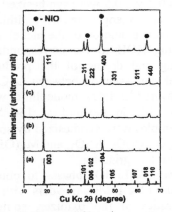

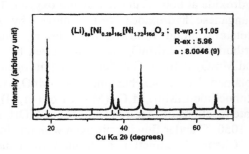

Fig. 2 X-ray diffraction patterns of (a) $LiNiO_2$, (b) $Li_{0.5}NiO_2$, (c) $Li_{0.5}NiO_2$ after heating at 150 °C, (d) $Li_{0.5}NiO_2$ after heating at 200 °C, and (e) $Li_{0.5}NiO_2$ after heating at 300 °C for 3 days.

Fig. 3 X-ray diffraction pattern, calculated profile, their difference, and peaks positions (Rietveld refinement results) for $Li_{0.5}NiO_2$ after heating at 200 °C. The cation distribution, R factors, and lattice parameter are also given.

consists of nickel ions both in the 16d and 16c sites of the spinel lattice (Fig. 3).

As discussed in the introduction, the transformation of the layered $(Li_{0.5})_{3a}(Ni)_{3b}O_2$ into the ideal cubic spinel phase $(Li)_{8a}[Ni_2]_{16d}O_4$ requires a migration of 25 % of the nickel ions from the nickel plane to the lithium plane. The presence of nickel ions in both the 16d and 16c sites in the 200 °C heated samples reveals that the migration of nickel ions is incomplete at the moderate temperatures of 200 °C. Further increase in temperature to achieve the required migration of 25% of the nickel ions and thereby the ideal spinel $(Li)_{8a}[Ni_2]_{16d}O_4$ results in a disproportionation of the phase to give NiO impurity as indicated by the X-ray data in Fig. 2e. The experiments thus reveal that it is difficult to access the ideal spinel phase with the cation distribution of $(Li)_{8a}[Ni_2]_{16d}O_4$.

Fig. 4 shows the evolution of X-ray diffraction patterns on heating at various temperatures the $Li_{0.5}CoO_2$ that was obtained by extracting lithium from the $LiCoO_2$ synthesized at 800 °C. The data reveal that a transition from rhombohedral to cubic symmetry does not occur in this system. The absence of such a transformation indicates that a migration of the cobalt ions from the cobalt planes to the lithium planes does not occur even on heating to 200 °C. Further

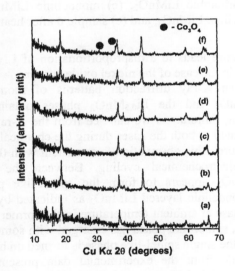

Fig. 4. X-ray diffraction patterns of (a) $Li_{0.5}CoO_2$ (obtained by delithiating $LiCoO_2$ that was synthesized at 800 °C), (b) $Li_{0.5}CoO_2$ after heating at 150 °C, (c) $Li_{0.5}CoO_2$ after heating at 200 °C, (d) $Li_{0.5}CoO_2$ after heating at 250 °C, (e) $Li_{0.5}CoO_2$ after heating at 300 °C, and (f) $Li_{0.5}CoO_2$ after heating at 400 °C for 3 days in air.

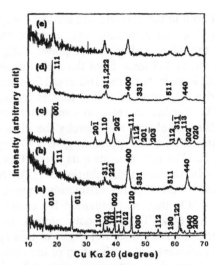

Fig. 5. X-ray diffraction patterns of (a) orthorhombic $LiMnO_2$, (b) $Li_{0.5}MnO_2$ obtained from orthorhombic $LiMnO_2$, (c) monoclinic $LiMnO_2$, (d) $Li_{0.5}MnO_2$ obtained from monoclinic $LiMnO_2$, and (e) sample d after heating at 150 °C.

increase in temperature leads to a disproportionation of $Li_{0.5}CoO_2$ into $LiCoO_2$ and Co_3O_4 as found in the case of the nickel oxide system.

Fig. 5 gives the X-ray diffraction patterns of monoclinic (layer) and orthorhombic $LiMnO_2$ and the $Li_{0.5}MnO_2$ phases obtained from them by chemically extracting lithium. As we can see from the X-ray data, a spinel-like phase is already formed in both the cases during the chemical delithiation process at ambient temperature. This observation is consistent with that found by several groups[3-5] during electrochemical cycling. Between the two systems, the orthorhombic $LiMnO_2$ appears to form the spinel-like phase more readily compared to the monoclinic layered $LiMnO_2$ as indicated by the formation of a single spinel-like phase at ambient temperature in the former case. In the case of monoclinic $LiMnO_2$, spinel-like phase has emerged with some unidentified peaks after extraction and the single spinel-like phase is formed on heating at 150 °C.

We can conclude from the experimental data presented above that the tendency of the layered $Li_{0.5}MO_2$ to transform to cubic spinel-like phase decreases in the order $Li_{0.5}MnO_2 > Li_{0.5}NiO_2 > Li_{0.5}CoO_2$. While a spinel-like phase is readily formed during lithium extraction for M = Mn, a mild temperature is needed for the case of M = Ni. On the other hand, it is difficult to obtain a spinel-like phase for M = Co even after heating at 200 °C.

Materials for Electrochemical Energy Conversion and Storage

Since the transformation of the layered phase to spinel-like phase involves the migration of the M^{n+} ions from the octahedral sites of one layer to the octahedral sites of the other layer via the tetrahedral sites, we need to consider the stability of the M^{n+} cations in the tetrahedral sites and the activation energies associated with the migration to understand the differences among the three systems. Although the exact calculation of the activation energy may involve a complex procedure, a qualitative understanding can be developed by considering the relative stability of the M^{n+} ions in the octahedral versus tetrahedral sites. Table I gives the crystal field stabilization energies (CFSE) for the octahedral and tetrahedral coordination along with the octahedral site stabilization energies (OSSE), which is the difference between the CFSE values for octahedral and tetrahedral coordinations, for various ions. For the M^{3+} ions (low spin (LS) configurations for Ni^{3+} and Co^{3+} and high spin (HS) configuration for Mn^{3+}), the OSSE values increase in the order $Mn^{3+} < Ni^{3+} < Co^{3+}$. A larger OSSE value for the Co^{3+} ions means that the migration via the tetrahedral sites will be difficult; therefore, the formation of a spinel-like phase becomes difficult for M = Co. On the other hand, a smaller OSSE value for Mn^{3+} makes the formation of spinel-like phase easier for M = Mn. With an intermediate value for OSSE, the M = Ni system needs mild heat.

With an aim to identify strategies that can suppress the cation migrations, we then investigated the effect of cationic substitutions on the formation of the spinel-like phases. Accordingly, we investigated the $Li_{0.5}Ni_{1-y}Co_yO_2$ ($0 \leq y \leq 1$), and $Li_{0.5}Ni_{1-y}Mn_yO_2$ ($0 \leq y \leq 0.5$) systems. An examination of X-ray diffraction patterns of $Li_{0.5}Ni_{0.5}Co_{0.5}O_2$ ($y = 0.5$) before and after heating at various temperatures reveals that the (018) and (110) reflections centered around $2\theta = 65°$ do not merge even after heating at 200 °C suggesting the difficulty of the formation of the spinel-like phase compared to that found with $Li_{0.5}NiO_2$. An

Table I. Crystal field stabilization energies (CFSE) and octahedral site stabilization energies (OSSE) of nickel, cobalt and manganese ions

Ion	Octahedral Coordination				Tetrahedral Coordination				OSSE (D_q)
	Configuration[a]		CFSE		Configuration		CFSE[b]		
	t_{2g}	e_g			e	t_2			
$Ni^{3+}:3d^7$	6	1	(LS)	-18	4	3	(LS)	-5.333	-12.667
$Ni^{4+}:3d^6$	6	0	(LS)	-24	3	3	(LS)	-2.667	-21.333
$Co^{3+}:3d^6$	6	0	(LS)	-24	3	3	(LS)	-2.667	-21.333
$Co^{4+}:3d^5$	5	0	(LS)	-20	2	3	(LS)	0.000	-20.000
$Mn^{3+}:3d^4$	3	1	(HS)	-6	2	2	(LS)	-1.778	-4.222
$Mn^{4+}:3d^3$	3	0	(HS)	-12	2	1	(LS)	-3.556	-8.444

[a] LS and HS refer, respectively, to low spin and high spin configurations.
[b] Obtained by assuming $\Delta_t = 4/9\Delta_o$; Δ_t and Δ_o refer, respectively, to tetrahedral and octahedral splittings.

analysis of the X-ray data for various y values in $Li_{0.5}Ni_{1-y}Co_yO_2$ indicates that the formation of spinel-like phases is suppressed with increasing Co content y.

An examination of the X-ray diffraction patterns of $Li_{0.5}Ni_{0.5}Mn_{0.5}O_2$ (y = 0.5) before and after heating at various temperatures also reveals that the (018) and (110) reflections centered around $2\theta = 65°$ do not merge even after heating at 150 °C suggesting the difficulty of formation of the spinel-like phase compared to that found with $Li_{0.5}MnO_2$ (Fig. 5) and $Li_{0.5}NiO_2$ (Fig. 2). However, the $Li_{0.5}Ni_{1-y}Mn_yO_2$ (y = 0.1, 0.3, and 0.5) system was found to consist of spinel-like and layer phases after heating at 200 °C unlike the $Li_{0.5}Ni_{0.5}Co_{0.5}O_2$ sample. The fraction of the spinel-like phase in the 200 °C heated $Li_{0.5}Ni_{1-y}Mn_yO_2$ samples was found to decrease with increasing y although the pure $Li_{0.5}MnO_2$ forms spinel-like phase easily at ambient temperature. It is possible that the layered phase may be richer in Ni and the spinel phase may be richer in Mn. The data reveal that the mixed cations $Ni_{1-y}Mn_y$ suppress the formation of spinel-like phases even though the individual systems consisting of the single cations M = Mn or M = Ni form spinel-like phases under similar conditions. This mixed cation strategy may be useful to develop structurally stable cathode hosts. The use of this mixed cation strategy is further explored with the manganese oxide system in the next section.

Improvement of the Structural Stability of Layered Lithium Manganese Oxide

Monoclinic $LiMnO_2$ that has a layer structure similar to $LiCoO_2$ can be accessed only by an ion exchange of $NaMnO_2$. However, substitution of small amount of Cr for Mn is known to give the monoclinic layered material by conventional high temperature reactions.[12] The Cr substitution is also known to improve the electrochemical property. Although the structural stability increases with increasing Cr content, it leads to a reduction in capacity. Therefore, it is desirable to develop a strategy that can offer good structural stability while maintaining the high capacity. Towards this objective, we have explored the use of mixed cation strategy discussed in the previous section. Accordingly, we made four samples with two different levels of substitution: $LiMn_{0.9}Cr_{0.1}O_2$, $LiMn_{0.9}Cr_{0.07}Al_{0.03}O_2$, $LiMn_{0.85}Cr_{0.15}O_2$, and $LiMn_{0.85}Cr_{0.07}Al_{0.06}Mg_{0.02}O_2$. While the first two samples have 10% cationic substitution, the last two samples have 15% cationic substitution. After extracting 50 % lithium, the four samples were found to maintain the layer structure without transforming to spinel-like phases. Also, the monoclinic distortion was found to be suppressed due to the oxidation of some Mn^{3+} to Mn^{4+} during lithium extraction. However, a decrease in the c/a ratio was observed on heating the phases at mild temperatures due to the migration of Mn^{3+}. Fig. 6 compares the variation of the % change in the c/a ratio with heat treatment temperature for the four samples. As the degree of substitution increases, the change in c/a ratio decreases. More importantly, with the same degree of 15% cationic substitution, the sample containing mixed cations show a

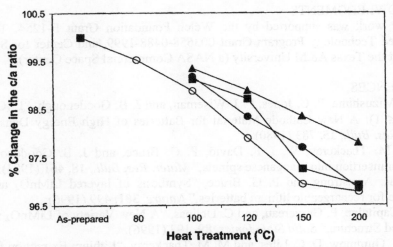

Fig. 6. % change in the *c/a* ratio of LiMn$_{0.9}$Cr$_{0.1}$O$_2$ (open circle), LiMn$_{0.9}$Cr$_{0.07}$Al$_{0.03}$O$_2$ (square), LiMn$_{0.85}$Cr$_{0.15}$O$_2$ (solid circle), and LiMn$_{0.85}$Cr$_{0.07}$Al$_{0.06}$Mg$_{0.02}$O$_2$ (triangle) on heating.

smaller change in the *c/a* ratio compared to the sample containing all Cr (15%). It appears that the presence of mixed cations is providing a better structural stability.

CONCLUSIONS

The ease of layered to spinel-like phase transition has been investigated by heating the chemically prepared Li$_{0.5}$MO$_2$ (M = Mn, Ni, and Co) compositions at mild temperatures. It is found that the ease of formation of spinel-like phase decreases in the order Li$_{0.5}$MnO$_2$ > Li$_{0.5}$NiO$_2$ > Li$_{0.5}$CoO$_2$, which could be related to the experimental observations of the electrochemical performance. This experimentally observed order has been explained on the basis of crystal field stabilization energies. Additionally, the presence of mixed cations in the transition metal layer is found to suppress the formation of spinel-like phases as illustrated with the systems Li$_{0.5}$Ni$_{1-y}$Co$_y$O$_2$ (0 ≤ y ≤ 1), Li$_{0.5}$Ni$_{1-y}$Mn$_y$O$_2$ (0 ≤ y ≤ 0.5), and Li$_{0.5}$Mn$_{1-y}$(Cr,Al,Mg)$_y$O$_2$. The mixed cations appear to perturb the cooperativity and thereby suppress the migration of cations from one layer to the other and the formation of spinel-like phase. The mixed cation strategy may prove effective to design structurally stable cathode hosts with high capacity.

ACKNOWLEDGMENTS

This work was supported by the Welch Foundation Grant F-1254, Texas Advanced Technology Program Grant 003658-0488-1999, and Center for Space Power at the Texas A&M University (a NASA Commercial Space Center).

REFERENCES

[1] K. Mizushima, P. C. Jones, P. J. Wiseman, and J. B. Goodenough, "Li_xCoO_2 ($0 < x \leq 1$): A New Cathode Material for Batteries of High Energy Density," *Mater. Res. Bull.*, **15**, 783 (1980).

[2] M. M. Thackeray, W. I. F. David, P. G. Bruce, and J. B. Goodenough, "Lithium insertion into manganese spinels," *Mater. Res. Bull.*, **18**, 461 (1983).

[3] A. R. Armstrong and P. G. Bruce, "Synthesis of layered $LiMnO_2$ as an electrode for rechargeable lithium batteries," *Nature*, **381**, 499 (1996).

[4] F. Capitaine, P. Gravereau, and C. Delmas, "A New Variety of $LiMnO_2$ with a Layered Structure," *Solid State Ionics*, **89**, 197 (1996).

[5] R. J. Gummow, D. C. Liles, and M. M. Thackeray, "Lithium Extraction from Orthorhombic Lithium Manganese Oxide amd the Phase-transformation to Spinel," *Mater. Res. Bull.*, **28**, 1249 (1993).

[6] T. Ohzuku, A. Ueda, and M. Nagayama, "Electrochemistry and Structural Chemistry of $LiNiO2$ (R3m) for 4 Volt Secondary Lithium Cells," *J. Electrochem. Soc.*, **140**, 1862 (1993).

[7] M. M. Thackeray, Y. Shao-Horn, A. J. Kahaian, K. D. Kepler, E. Skinner, J. T. Vaughey, and S. A. Hackney, "Structural Fatigue in Spinel Electrodes in High Voltage (4 V) $Li/Li_xMn_2O_4$ Cells," *Electrochem. Solid State Lett.*, **1**, 7 (1998).

[8] S. J. Wen, T. J. Richardson, L. Ma, K. A. Striebel, P. N. Ross, and E. J. Cairns, "FTIR Spectroscopy of Metal Oxide Insertion Electrodes: A New Diagnostic Tool for Analysis of Capacity Fading in Secondary Li/LiMn2O4 Cells," *J. Electrochem. Soc.*, **143**, L136 (1996).

[9] R. V. Chebiam, F. Prado, and A. Manthiram, "Structural Instability of Delithiated $Li_{1-x}Ni_{1-y}Co_yO_2$ Cathodes," *J. Electrochem. Soc.* **148**, A49 (2001).

[10] G. Ceder, and A. Vander Ven, "Phase diagrams of lithium transition metal oxides: investigations from first principles," *Electrochimica Acta*, **45**, 131 (1999)

[11] M.G.S.R. Thomas, W.I.F. David, and J.B. Goodenough, "Synthesis and Structural Characterization of the Normal Spinel $LiNi_2O_4$," *Mat. Res. Bull.*, **20**, 1137 (1985)

[12] I. J. Davidson, R. S. McMillan, and J. J. Murray, "Rechargeable Cathodes Based on $Li_2Cr_xMn_{2-x}O_4$," *J. Power Sources*, **54**, 205 (1995).

MODELING AND DESIGN OF INTERMETALLIC ELECTRODES FOR LITHIUM BATTERIES

R. Benedek, J. T. Vaughey,
and M. M. Thackeray
Argonne National Laboratory
Argonne, Illinois 60439

ABSTRACT

Intermetallic compounds are being investigated as a possible alternative to carbonaceous materials as anodes in lithium batteries. Upon lithium insertion, intermetallic compounds undergo two basic types of reactions, addition (to form a ternary compound) and displacement (in which Li substitutes for the less active element, which is extruded). The displacement reaction is unique to compounds and does not occur in pure metal electrodes. Theoretical modeling of the thermodynamics and kinetics of such reactions is discussed.

INTRODUCTION

Although graphitic carbon is the most widely used anode material in lithium batteries [1], safety concerns associated with the plating of Li on the anode during overcharging have stimulated investigations of other candidate materials, including metals and intermetallic compounds [2], which eliminate the plating problem at the cost of a reduction in cell potential. The search for suitable metallic electrodes has been guided primarily by the concept of mixing active and inactive components. Relatively little guidance has been provided by theoretical considerations or modeling. The purpose of this presentation is to discuss, from the standpoint of theory, the behavior of metallic electrodes during lithium insertion and extraction.

Pure polyvalent metals (and semiconductors) such as Al, Si, Sn, Pb, and Sb each form characteristic compounds with Li. The theoretical capacity for the first cycle is typically large. The volume expansion that accompanies Li uptake, however, leads to rupture of the specimen, which either reduces capacity or causes failure upon subsequent cycling. This behavior has motivated consideration of either active-inactive composites [3] or intermetallic compounds. The inactive component in composites is intended to buffer the volume-expansion-induced stresses in the active portion of the system by providing a ductile medium into which expansion can occur.

Composites are an area of active study to which we will refer only tangentially; instead we focus the present discussion on intermetallic compounds. It is worth emphasizing, however, that for both composites and intermetallic systems, the evolution of the inactive component with cycling is critical to the problem of capacity retention.

When measured in the dimensionless form, $d(lnV)/d(x_{Li})$, the volume expansion that accompanies lithiation in intermetallic compounds (of composition Li_xAB_y, say) is typically less than in pure metals because the ratio of Li to host atoms is less at a given x. This is offset, however, by the greater brittleness of intermetallics associated with their directional bonding, with the consequence that lithiation-induced rupture is still a significant problem in intermetallics, as it is with pure-metal electrodes.

Because of the extra compositional degrees of freedom in compounds, solid-state reactions with Li are possible that do not have any analog in reactions with pure materials. In particular, the *displacement* reaction, in which Li displaces the less active component of the compound, opens new possibilities. If the original compound and the product lithium compound are structurally compatible, the problem of rupture induced by volume expansion may, at least, be reduced relative to that in single-component electrodes. Considerable attention will be given to such displacement reactions in our discussion below.

In recent years, atomic-scale modeling, particularly by first-principles density functional theory [4] and molecular dynamics [5], have been applied to oxide electrodes. Kinetic models of time-dependent properties, based on rate-theory formulations, have also been developed [6]. As mentioned above, relatively little modeling has been applied to intermetallic electrodes. First-principles calculations can predict electrochemical potentials for Li reactions with intermetallic compounds with reasonable accuracy, as we describe below. The performance of the anode material, however, particularly capacity retention, is dominated by kinetic factors. Modeling the kinetics is difficult because the key processes or parameters have not been identified, and because of the wide range of time and length scales involved. While specific aspects of kinetics can be addressed, e.g., by first-principles calculations, modeling the kinetics in a comprehensive way is not possible at present.

In the remainder of this paper, we first address the thermodynamics of Li reactions with intermetallic compounds. This is followed by a discussion of kinetics, for which progress has unfortunately been minimal, in spite of its importance. Although theory cannot describe in a comprehensive way electrochemical process of lithium cycling in intermetallics, it may help in suggesting promising families of compounds for investigation, which is the subject of the latter part of the paper. Three general criteria are proposed that promote efficient Li reaction with intermetallics, and their relation to theory is discussed.

THERMODYNAMICS
Solubility of Li

The amount of Li that may be inserted without chemical reaction into intermetallics is considerably less than, for example, in transition metal oxides or graphite. The solubility of a given point defect, impurity, or complex in an intermetallic host is given essentially by [7]

$$s = exp(-\Delta g_d / kT), \tag{1}$$

where Δg_d is the formation free energy of the defect. In intermetallic compounds of Li, unlike in many transition metal oxides, Li is a chemically bonded structural component, rather than a weakly bonded intercalant. The consequence is that the energy cost to dissociate the Li-compound precipitate and create a Li-atom solute in the original intermetallic host, $\Delta g_d >> kT$, and only trace solubilities of Li exist in such intermetallic compounds. Supersaturated solutions of Li with concentrations $c_{Li} > s_{Li}$ are possible, in principle, if the barriers to transformation are high, but in our further discussion we assume only trace concentrations of Li occur in a given intermetallic host. Therefore, after a small amount of Li insertion, solid state chemical reactions commence.

Solid State Reactions of Li

In general, Li reacts with pure metals (or semiconductors) to form Li_xA compounds. For many elements, e.g. Sn [7], several such compounds (with different values of x) exist. The reaction

$$xLi + A \rightarrow Li_xA \tag{I}$$

is referred to as a *reconstitution* or *addition* reaction. Analogous addition reactions

$$xLi + AB_y \rightarrow Li_xAB_y \tag{II}$$

occur for compounds AB_y. In general, the product and the host have markedly different crystal structures. In intermetallic compounds, another type of reaction is possible,

$$x\text{Li} + AB_y \rightarrow \text{Li}_x B_y + A, \tag{III}$$

which is referred to as a *displacement* reaction. A third, *hybrid*, type of reaction is possible in which both addition and displacement occur, i.e., the component A is only partially displaced, and a ternary product and pure A are produced in the reaction. In reactions II and III, we regard A as the less active and B as the more active element.

Electrochemical Potentials

In a cell that consists of a metallic Li anode and a metallic cathode, the electrochemical potential is defined as

$$V = | \mu_{\text{Li}}(\text{product}) - \mu_{\text{Li}}(\text{Li}) |, \tag{2}$$

where μ_{Li} is the chemical potential of Li either in the product phase or in metallic Li.

In the case of a pure metal, cf. reaction I, the cell potential, neglecting kinetic effects, is simply related to the formation energy ΔG_F of compound $\text{Li}_x A$:

$$V_I = \Delta G_F(\text{Li}_x A)/x. \tag{3}$$

In the case of the addition reaction II,

$$V_{II} = \Delta G_F(\text{Li}_x AB_y)/x - \Delta G_F(AB_y)/x. \tag{4}$$

Similar expressions are readily obtained for displacement and addition-displacement reactions, but are omitted here for the sake of brevity. In the case of binary intermetallic systems that don't form ternaries with Li, only displacement reactions come into consideration.

If we exclude for the moment hybrid addition-displacement reactions, three types of behavior are possible for particular intermetallic compounds: (i) addition reaction followed by displacement reaction, (ii) addition reaction(s) only, and (iii) displacement reaction only. Since relatively few ternary intermetallic compounds with Li exist, category (iii) is the most common.

Materials for Electrochemical Energy Conversion and Storage

Electrochemical Potential Prediction and Comparison with Experiment

For pure metals, electrochemical potential predictions based on eq. (2) may be obtained from calorimetric measurements of compound formation energies [8], where available. Alternatively, calculations may be performed using first-principles methods. In the results shown below, first-principles local-density-functional-theory calculations are performed using the VASP code [9,10]. The generalized-gradient-approximation (GGA) correction is applied in these calculations. In each case, internal atomic coordinates and lattice parameters are relaxed to minimize internal energy. The calculations strictly apply only to 0 K, but free energies of compounds at room temperature differ only slightly from those at lower temperatures, relative to typical cell voltages.

In Fig. 1 are plotted theoretical and measured electrochemical potentials corresponding to the first chemical reactions upon Li insertion in selected compounds and pure materials. The pure materials include Al [3] and Sb [3], and the compounds include Al_2Cu [12], MnSb [13], InSb [14], Sb_2V [12], and Cu_2Sb [15]. The calculated values (abscissa) are obtained with the VASP code, whereas the ordinates are taken from measured plateau values that correspond to the first lithiation cycle. Typically, the plateaus obtained for intermetallic compounds are more sloping than those for pure metals. An example is shown in Fig. 2, measured for InSb [16]. Measurements with EXAFS on the first insertion cycle indicate that the displacement reaction, of the type given in reaction III, commences at about 0.75 eV and continues to about 0.6 eV, on the ball-milled specimen. This result indicates that the start of the reaction corresponds roughly to the point of inflection that occurs between the initial transient (approximately the first 3 h) and the relatively flat part of the curve (from 3 to 10 h). The experimental values plotted in Fig. 2 are estimated in an analogous way.

The theoretical results plotted in Fig. 2 tend to overestimate slightly the cell voltage, as compared to experiment. This is in contrast to results based on local-density-functional theory for transition-metal oxides, in which the voltage is systematically underestimated [4]. In the case of the oxides, the underestimation is attributed in large part to the large cohesive energy predicted for Li , which is reduced somewhat by the GGA. In the case of metallic electrodes, the stress imposed on the product phase, which grows in the matrix of the host system, may shift its free energy, relative to that in equilibrium, and contribute to the observed trend of overestimated cell voltages for metallic systems.

KINETICS

Several factors influence the kinetics of the insertion-extraction cycle for intermetallic compounds, including (a) the diffusion of Li as well as one or more of the host components, (b) the nucleation of the product phase (a ternary compound in the case of an addition reaction, and a binary Li compound in the case of a displacement reaction), (c) the characteristics of the reaction-front interface, and (d) internal stress accommodation. Because of their complex interplay, and most important, their disparate length and time scales, these phenomena have thus far not yielded to any comprehensive theoretical treatment.

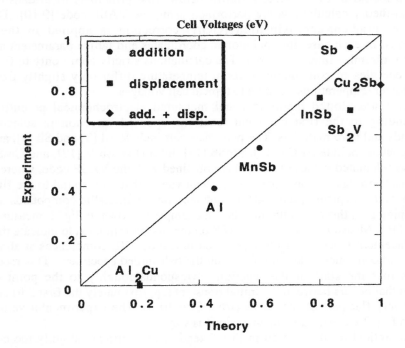

Fig. 1. Experimental vs. predicted cell voltages for Li-induced reactions with several intermetallic compounds. Theoretical values were obtained with the VASP code [10,11], using the generalized gradient approximation (GGA). The experimental atomic structure, where known, was employed as the starting configuration, and lattice parameter and internal coordinates were optimized. In the case of pure Mn, an approximate antiferromagnetic structure for the 58-atom unit-cell was assumed. In the case of Al_2Cu, no electrochemical reaction was observed experimentally [12]. Calculations were performed for the ternary $LiAl_2Cu$, based on a hypothetical hexagonal structure. Similar voltages were predicted for addition and displacement reactions for this system. Experimental values were obtained as discussed in reference to Fig. 2.

Schmalzried [17] emphasizes the complexity of the reaction-interface morphology as an obstacle to the quantitative interpretation of chemically-driven displacement

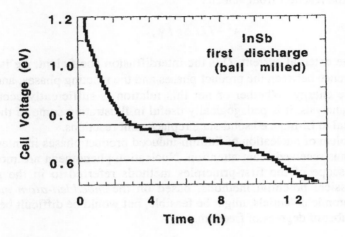

Fig. 2. Electrochemical potential for the first discharge cycleof ball-milled InSb [16]. The horizontal lines indicated the voltages between which In displacement occurs, as measured by extended x-ray absorption fine structure (EXAFS) analysis.

reactions. The morphology of the interface in displacement reactions was discussed as early as the 1930's by C. Wagner. The interleaved structure of the laminate in an electrochemical cell makes experimental analysis of the electrode particle microstructures particularly challenging.

Since the introduction of Li into metallic systems generally causes substantial volume expansion, stress accommodation is a critical issue for maintaining the integrity of a battery system. Stress accommodation may occur, for example, by enhanced diffusion, plastic deformation, or rupture. Rupture poses a severe problem for capacity retention. For active-inactive composite systems, fine particle sizes minimize or eliminate the problem of rupture [18].

Kinetic models for addition [19] and displacement [20] reactions in the steady-state regime have been developed, particularly for oxides. The electrochemical reactions in lithium batteries differ from such applications in that the Li flux is determined by a current that is set externally; furthermore, Li batteries are typically operated near room temperature, whereas most chemically driven solid-state reactions occur at elevated temperatures.

A simplified analytical treatment of a Li-induced displacement reaction in the steady-state, diffusion-limited regime was presented recently by Hackney [21]. Imposing the *maximum-reaction-rate* condition, he obtained the following relation for the reaction-front velocity v:

$$v^2 \sim I D \Delta g / \gamma, \qquad (5)$$

where I is the external current, D is the interdiffusion coefficient, Δg is the free energy difference between the product phases and the reacting phases, and γ is the interface free energy. Whether or not this relation is sufficiently accurate for quantitative analysis, it is pedagogically useful in illustrating some of the factors that are critical to Li-induced solid-state replacement reactions.

The problem of nucleation of lithium-induced product phases in intermetallic compounds has received scant attention. Nucleation phenomena are most likely beyond the scope of the first-principles methods referred to in the previous section. Classical potential methods, based on the *embedded-atom method* or similar interatomic potentials, might be feasible, but would be difficult because of the huge number of degrees of freedom.

Divergence between Break-in and Subsequent Cycles

As mentioned above, addition and displacement reactions, (II and III, respectively), are the primary transformations that occur upon lithium insertion into an intermetallic compound host. The displacement reaction is by far the more common of the two, since ternary Li compounds only occur for a minority of systems.

Most (although not necessarily all [12]) intermetallic compounds that don't undergo addition reactions with Li exhibit displacement reactions at room temperature. Upon charging (lithium extraction), however, the typical behavior is that the reverse reaction does not occur, and a composite of active and inactive phases remains, in place of the original compound. This is found, for example, for silicides [22], various antimonides [23], and alloys with the Al_2Cu structure [12]. The compounds that undergo a more or less irreversible displacement reaction during the first cycle, therefore, provide a procedure for synthesizing composite systems, which it is not our intention to discuss here. It is instead the intermetallic compounds for which the displacement reaction is at least substantially reversible that gives intermetallics a significance independent of composite systems.

Recent work by the Argonne National Laboratory (ANL) battery group has focused on systems that have appreciable reversibility, including Cu_6Sn_5 [24], InSb [25], and Cu_2Sb [15]. The reversibility is demonstrated by in-situ x-ray diffraction. The behavior of the Cu-based compounds with Sn and Sb is complicated by the fact that addition reactions occur prior to the displacement reactions. The reversibility of these systems is substantial but not complete. In the case of InSb, about 80% of the In was displaced in the first cycle, and about half of that reentered the InSb matrix upon charging [16].

SCREENING CRITERIA

The existing theory of intermetallic-compound electrodes, outlined in the previous section, is unable to predict kinetically determined properties, such as capacity retention, that are critical to battery performance. It is, therefore, unable by itself to assess the suitability of a particular electrode material for practical or commercial use. Against this background, one may ask whether a set of theoretically motivated criteria can at least identify "promising" candidate materials for experimental study. In this section we propose a set of three criteria and discuss their consequences. The criteria relate only to structural features and do not address practical constraints of cost, toxicity, etc.

(i) Absence of addition reactions. Addition reactions unavoidably lead to volume expansion and internal stresses that are likely to cause rupture. In displacement reactions, on the other hand, the diffusion of the inactive species of the compound provides an additional mechanism for relieving the stress induced by Li insertion. Displacement reactions, therefore, appear more promising than addition reactions for preserving the structural integrity of the electrode. We note, however, that two of the three reversible compounds mentioned above undergo addition as well as displacement reactions.

(ii) Crystallographic compatibility. The ability of the original intermetallic compound to transform to a Li compound in a displacement reaction is expected to be promoted if the structures of the host and product are as similar as possible. In particular, we require that the active element sublattice possess the same structure (for example, cubic close packed) in the host compound and in the product compound; examples will be given below. We refer to this sublattice as *invariant*.

(iii) minimum volume misfit. The reaction should be further facilitated if the volume misfit between host and product is minimized, i.e., if the lattice constants of the invariant sublattice in host and product phases are as similar as possible.

In spite of their limited scope, these criteria are already highly restrictive. A search of listed intermetallic compounds [26] reveals only a few compounds (all belonging to the III-V class) that satisfy criterion (ii). In Fig. 3 are plotted the lattice constants for compounds of cubic antimonides and bismuthides with trivalent elements (group IIIA, IIIB, or rare earth elements) as a function of the trivalent element atomic number. The antimonides adopt the zinc blende structure, whereas the bismuthides exhibit the rock salt structure. In displacement reactions

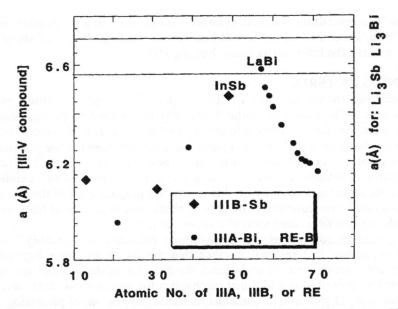

Fig.3 Lattice constant data [26] for compounds of trivalent (group IIIA, IIIB, and rare-earth [RE]) elements with Sb or Bi. The horizontal lines represent the lattice constants of Li_3Sb and Li_3Bi. The compounds InSb and LaBi give the smallest misfit (<2%) relative to the displacement product phase Li_3Sb or Li_3Bi.

for these host materials, the product phases are Li_3Sb and Li_3Bi, both of which adopt the cubic BiF_3 structure. The plot illustrates that criterion (iii) is best satisfied by the compounds InSb and LaBi; the lattice constant mismatch for both compounds (with Li_3Sb and Li_3Bi respectively), is less than about 2%. Extensive experimental work has been performed on InSb by the ANL group [25,27]; we are unaware of any work on LaBi. Since LaBi forms a ternary, Li_3LaBi_2, it strictly violates the first criterion, although this may not pose a serious problem in practice.

The suitability of InSb as an electrode material has been questioned by the Dalhousie University group [28]. Satisfaction of the three postulated criteria, of course, does not guarantee favorable electrode performance. We have, however, continued to investigate InSb in order to evaluate the design strategy espoused here. Whether or not InSb turns out to be a useful electrode material, it is unusual in exhibiting a somewhat reversible displacement reaction with Li at ambient temperature. We believe that this behavior is, in part, a consequence of its meeting the three criteria.

　Materials for Electrochemical Energy Conversion and Storage

As mentioned above, Cu_6Sn_5 and Cu_2Sb, as well as InSb, also show reversibility. Unlike InSb, these Cu compounds undergo both addition and displacement reactions, and thus violate criterion (i). Criterion (ii) is satisfied by displacement reactions that occur for Li_2CuSn and Li_2CuSb, respectively, but the misfit is larger than for InSb.

Implications of *Screening Criteria* for Kinetics

We now discuss how the proposed screening criteria relate to the kinetics of intermetallic insertion electrodes. In general, criteria (ii) and (iii) are expected to facilitate reaction-front kinetics. For example, they should yield relatively small reaction-front interface energies, γ, which in turn relate to transformation velocities [cf. eq. (5)] and critical nucleus sizes. Experience with oxide reactions [29] suggests that a small misfit (criterion 3) should promote growth of the product in the reaction-limited phase of the growth process. Criterion (iii) is also expected to minimize the cracking induced by volume mismatch, assuming the displaced inactive component (In in InSb) can be efficiently transported to the particle surface. Finally, the small mismatch is expected to promote reversibility of the displacement reaction. Reversibility is crucial for capacity and capacity retention.

The proposed criteria thus may yield favorable reaction-front properties and reduce internal stress; however, they do not address the diffusion kinetics of the displaced component. In order for a displacement reaction to proceed, the displaced component must be transported efficiently to the surface. The detailed mechanisms associated with this process are specific to the particular material in question, for example, InSb.

Unfortunately, the kinetics of In displacement by Li in InSb are poorly understood. Scanning electron microscopy observations [30] indicate that the displaced In is extruded in the form of whiskers; whiskers have not been observed for Cu_2Sb and Cu_6Sn_5, however. Whisker growth is a well-known phenomenon [31], and some qualitative features, such as the importance of stress, are generally accepted. Quantitative models are lacking, however, and experience with whisker growth in an electrochemical context is limited. The lack of understanding of the whisker-growth process is an obstacle at this time.

CONCLUSIONS

Intermetallic compounds undergo addition, displacement, or combined addition-displacement reactions with Li; a sequence of reactions, for example addition followed by displacement, occurs in some systems. The displacement reaction, which has no counterpart in pure-metal electrodes, is a unique feature of intermetallic compounds. Most intermetallic compounds for which displacement reactions occur, however, stay essentially phase-separated upon subsequent cycling. Some compounds for which the insertion reactions are at least partially reversible are InSb, Cu_2Sb, and Cu_6Sn_5.

First-principles density-functional theory calculations predict cell voltages for addition and displacement reactions of intermetallic compounds in close agreement with experiment. Existing models of the kinetics of lithium reactions with intermetallic compounds, however, are not presently able to predict performance of real materials.

We proposed three criteria for identifying promising intermetallic compounds: absence of addition reactions, crystallographic compatibility, and minimum volume misfit. Specifically, a search of existing compounds was made for those that exhibit displacement reactions for which the product and host phases are structurally compatible and have small misfit. The two systems that best satisfy the criteria are InSb and LaBi, the first of which has been studied experimentally. An important aspect of displacement reactions that is poorly understood is the diffusion of the displaced (less active) component of the intermetallic compound. Additional work, both experimental and theoretical, is required to understand better the factors that determine capacity and capacity retention.

ACKNOWLEDGMENTS

This work was supported at Argonne National Laboratory by the Chemical Sciences Division of the Office of Basic Energy Sciences of the U.S. Department of Energy, under contract no. W31-109-Eng-38. Most of the computational work was performed at the National Energy Research Supercomputer Center.

REFERENCES

[1] J. Yamaki, M. Egashira, and S. Okada, "Potential and Thermodynamics of Graphite Anodes in Li-ion Cells," *J. Electrochem. Soc.* **147**, 460-465 (2000).
[2] R. A. Huggins, "Lithium Alloy Anodes", pp. 359-381, in *Handbook of Battery Materials*, edited by J. O. Besenhard, Wiley, New York (1999), and other articles in this volume.
[3] M. Winter and J. O. Besenhard, "Electrochemical Lithiation of Tin and Tin-Based Intermetallics and Composites," *Electrochim. Acta* **45**, 31-50 (1999).
[4] For example, M. K. Aydinol, A. F. Kohan, G. Ceder, K. Cho, and J. Joannopoulos, "Ab Initio Study of Lithium Intercalation in Metal Oxides and metal Dichalcogenides,",*Phys. Rev.* B **56**, 1354 (1997).
[5] B. Ammundsen, J. Roziere, and M. S. Islam, "Atomistic Simulation Studies of Lithium and Proton Insertion in Spinel Lithium Manganates," *J. Phys. Chem.* B **101**, 8156-8163 (1997).
[6] L. Song and J. W. Evans, "Electrochemical-Thermal Model of Lithium Polymer Batteries," *J. Electrochem. Soc.* **147**, 2086-2095 (2000), and references therein.

[7]C. G. Van de Walle, D. B. Laks, G. F. Neumark, and S. T. Pantelides, "First-Principles Calculations of Solubilities and Doping Limits: Li, Na, and N in ZnSe,",*Phys. Rev.* B **47**, 9425-9433 (1993).

[8]I. A. Courtney, J. S. Tse, O. Mao, J. Hafner, J. R. Dahn, "Ab Initio calculation of the lithium-tin voltage profile," *Phys. Rev.* B **58**, 15583-15588 (1998).

[9]*Selected Values of Thermodynamic Properties of Binary Alloys*, A. Hultgren, ed., Am. Society for Metals, Metals Park, Ohio, (1973).

[10]G. Kresse and J. Furthmuller, "Efficiency of ab-initio Total-Energy Calculations for Metals and Semiconductors Using a Plane-Wave Basis Set," *Comput. Mat. Sci.* **6**, 15 (1996).

[11]G. Kresse and J. Furthmuller, "Efficient Iterative Schemes for Ab Initio Total-Energy Calculations Using a Plane-Wave Basis Set," Phys. Rev. B **54**, 11169-11186 (1996).

[12]D. Larcher, L. Y. Beaulieu, O. Mao, A. E. George, and J. R. Dahn, "Study of the Reaction of Lithium with Isostructural A_2B and Various Al_xB Alloys," *J. Electrochem. Soc.* **147**, 1703-1708 (2000).

[13]J. T. Vaughey, et al., unpublished (2000).

[14]J. T. Vaughey, J. O'Hara, and M. M. Thackeray, "Intermetallic Insertion Electrodes with a Zinc-Blende-Type Structure for Li Batteries: A study of Li_xInSb (0<x<3)," *Electrochem. Solid-State Lett.* **3**, 13-16 (2000).

[15]L. Fransson, et al., unpublished (2000).

[16]A. J. Kropf, H. Tostmann, C. S. Johnson, J. T. Vaughey, and M. M. Thackeray, "An In Situ X-Ray Absorption Spectroscopy Study of InSb Electrodes in Lithium Batteries," submitted to *Electrochem. Commun.* (2001).

[17]H. Schmalzried, *Chemical Kinetics of Solids* (VCH, Weinheim, 1995), p. 155.

[18]J. Besenhard, unpublished; R. A. Huggins, unpublished.

[19]J.-Y. Huh, T. Y. Tan, and U. Goesele, "Model of Partitioning of Point Defect Species during Precipitation of a Misfitting Compound in Czochralski Silicon," *J. Appl. Phys.* **77**, 5563-5571 (1995).

[20]R. A. Rapp, A. Ezis, G. J. Yurek, "Displacement Reactions in the Solid State", *Met. Trans.* **4**, 1283-1292 (1973); G. J. Yurek, R. A. Rapp, and J. P. Hirth, "Kinetics of the Displacement Reaction between Iron and Cu_2O," *Met. Trans.* **4**, 1293-1300 (1973).

[21]S. R. Hackney, "Multi-Phase Transformations in Intermetallic Compounds for Li-ion Batteries," unpublished work, 2001.

[22]G. X. Wang, L. Sun, D. H. Bradhurst, S. Zhong, S. X. Dou, H. K. Liu, "Nanocrystalline NiSi Alloy as an Anode Material for Lithium-ion Batteries," *J. of Alloys and Compounds* **306**, 249-252 (2000); "Innovative Nanosize Lithium Storage Alloys with Silica as Active Centre," *J. Power Sources* **88**, 278-281 (2000).

[23]X. B. Zhao, G. S. Cao, C. P. Lv, L. J. Zhang, S. H. Hu, T. J. Zhu, B. C. Zhou, "Electrochemical Properties of Some Sb or Te Based Alloys for Candidate Anode Materials of Lithium-Ion Batteries," *J. Alloys and Compounds* **315**, 265-269 (2000).

[24]M. M. Thackeray, J. T. Vaughey, A. J. Kahaian, K. D. Kepler and R. Benedek, "Intermetallic Insertion Electrodes Derived from NiAs, Ni_2In, and Li_2CuSn-type structures for lithium-ion batteries," *Electrochem. Commun.* **1**, 111-115 (1999). and unpublished work.

[25]J. T. Vaughey, J. O'Hara, and M. M. Thackeray, "Intermetallic Insertion Electrodes with a Zinc Blende-Type Structure for Li Batteries: A Study of LiInSb (0<x<3)," *Electrochem. Solid-State Lett.* **3**, 13-16 (2000).

[26]P. Villars and L.D. Calvert, *Pearson's Handbook of Crystallographic Data for Intermetallic Phases*, second edition, ASM, Materials Park, OH (1991).

[27]C. S. Johnson, J. T. Vaughey, M. M. Thackeray, T. Sarakonsri, S. A. Hackney, L. Fransson, K. Edstrom, and J. O. Thomas, "Electrochemistry and In-situ X-ray Diffraction of InSb in Lithium Batteries," *Electrochem. Commun.* **2**, 595-600 (2000).

[28]K. C. Hewitt, L. Y. Beaulieu, and J. R. Dahn, "InSb Is Not a Good Li Insertion Host," abstract submitted to Electrochemical Society Meeting, Phoenix, Oct. 2000.

[29]H. Sieber, D. Hesse, and P. Werner, "Misfit Accomodation Mechanisms at Moving Reaction Fronts during Topotaxial Spinel-Forming Thin-Film Solid-State Reactions: A High-Resolution Transmission Electron Microscopy Study of Five Spinels of Different Misfits," *Phil. Mag.* A **75**, 889-908 (1997); H. Sieber, P. Werner and D. Hesse, "The Atomic Structure of the Reaction Front as a Function of the Kinetic Regime of a Spinel-Forming Solid-State Reaction," *Phil. Mag.* A **75**, 909-924 (1997).

[30]S. R. Hackney, unpublished, 2000.

[31]See, for example, K. N. Tu, "Irreversible Process of Spontaneous Whisker Growth in Bimetallic Cu-Sn Thin-Film Reactions," *Phys. Rev.* B **49**, 2030-2034 (1994).

Il-seok Kim[a] and Prashant N. Kumta[a]

[a]Carnegie Mellon University, Pittsburgh, Pennsylvania 15213

G. E. Blomgren[a,*]

*Blomgren consulting Services Ltd., 1554 Clarence Avenue, Lakewood, Ohio 44107

ABSTRACT

Silicon based nanocomposites containing TiN were synthesized by high-energy mechanical milling (HEMM). Mechanical milling leads to very fine amorphous silicon particles distributed homogeneously inside the TiN matrix. The Si/TiN nanocomposites synthesized using different experimental conditions were evaluated for their electrochemical properties. Results indicate that silicon in the composite alloys and de-alloys with lithium during cycling, while TiN remains inactive providing structural stability. The composite containing 33.3 mol% silicon obtained after milling for 12 h exhibited a high capacity, ≈ 300 mAh/g with good capacity retention reflective of the good phase and micro-structural stability as verified by XRD and SEM analyses. Conventional and high-resolution electron microscopy coupled with electron energy-loss spectroscopy conducted on the composites validated the existence of amorphous silicon in a nanocrystalline TiN matrix.

INTRODUCTION

Graphite has so far been the customary anode material for lithium ion batteries with a theoretical capacity of 372 mAh/g or volumetric capacity of 830 Ah/L.[1] During the last few years, however, much of the research efforts have been directed towards identifying alternative anode materials. Fuji demonstrated the potential of nanostructured materials for lithium-ion anodes in a series of patents related to tin oxide.[2] Electrochemical insertion of lithium leads to the *in situ* generation of nanosized tin clusters in a matrix comprising of an electrochemically inactive oxide glass. The system is indeed promising demonstrating an initial reversible capacity of ≈ 600 mAh/g. A major limitation of the system is however the irreversible loss of lithium consumed in the electrochemical reduction of tin oxide. Nevertheless, these initial results helped to demonstrate the possibility of minimizing or perhaps even eliminating the large volume-induced strain during cycling, characteristic of lithium containing Zintl phases by the generation of active-inactive nanostructured composites. The validity of these concepts has been further successfully demonstrated in the Sn-Fe-C, Cu-Sn and other intermetallic systems.[3-6]

In order to preserve and stabilize the original morphological state of the anode and thereby attain good electrochemical properties, various material systems have been analyzed to minimize the mechanical stress caused by the large phase transition induced volumetric changes experienced by the active phase. Most of the current studies on anode materials other than carbon have focused on creating a composite microstructure comprising an inactive host matrix containing a finely dispersed interconnected active phase.[2-6] Although these systems are promising, there are problems related to either irreversible loss, capacity and/or cyclability. These results make it very essential to focus on approaches to improve further the demonstrated concept of active-inactive composites. This would require the identification of suitable materials systems and furthermore, the use of an appropriate approach for synthesizing the nanocrystalline composite.

The technique of mechanical alloying is known for its ability to generate nanocomposite structures of metals, ceramics, semiconductors and even polymers. In this study, we have exploited this approach to improve upon the active-inactive concept further by exploring the generation of nanocrystalline composites in the Si/TiN system for use as an anode material. The composite was synthesized by using the technique of high-energy mechanical milling (HEMM). TiN is a well-known technological material used in several applications.[7-9] Due to its good electrical conductivity, high surface area TiN is also recently attracting interest for supercapacitor applications.[10, 11] However, to the best of our knowledge, there have been no reports to date on its use as an electrode in lithium ion batteries. The basic premise of this work is therefore to demonstrate the usefulness of TiN as an electrochemically inactive matrix in the presence of an electrochemically active silicon phase. This is because of the well-known electrical and mechanical properties of TiN such as electrical conductivity, mechanical strength, combined with electrochemical inertness to lithium (potential range $0.02 \rightarrow 1.2$ V)[12] and chemical inertness to both lithium and silicon. However, these favorable attributes of TiN with respect to lithium and silicon can only be exploited if a suitable process for synthesizing such composites can be identified. In this context, the technique of HEMM is indeed very promising due to its proven ability to generate amorphous, metastable, and nanophase structures.[13] Initial results on this system were reported earlier by us.[14] The present paper therefore focuses on using HEMM for generating novel nanocomposites in the Si-TiN system for use as anodes in Li-ion rechargeable batteries. Experimental studies and results of the structural and electrochemical analyses are presented indicating its promising nature. In addition, electron microscopy results validating the formation of nanocomposites are also provided.

EXPERIMENTAL

Nanocomposites of silicon and TiN were prepared using a SPEX-8000 high-energy mechanical mill. Commercially obtained elemental powder of silicon (Aldrich, 99%) and TiN (Aldrich, 99%)) were used as starting materials. Stoichiometric amounts of the powder were weighed and loaded into a hardened steel vial containing hardened steel balls. All the processes prior to milling were conducted inside the glovebox (VAC Atmospheres, Hawthorn, CA) and the vial was firmly sealed to prevent and minimize any oxidation of the silicon and titanium nitride.

In order to evaluate the electrochemical characteristics, electrodes were fabricated using the as-milled powder by mixing the composition containing 87.1 wt.% active powder and 7.3 wt.% acetylene black. A solution comprising 5.6 wt.% polyvinylidene fluoride (PVDF) in 1-methyl-2-pyrrolidinone (NMP) was added to the mixture. The as-prepared solution was coated onto a Cu foil. A hockey puck cell design was used employing lithium foil as an anode and 1 M $LiPF_6$ in EC/DMC (2:1) as the electrolyte. All the batteries tested in this study were cycled for 20 cycles in the voltage range from 0.02~1.2 V employing a current density of 0.25 mA/cm^2 and a rest period of 60 s between the charge/discharge cycles using a potentiostat (Arbin electrochemical instrument). The phases present in the as-milled powders and the cycled electrode were analyzed using x-ray diffraction (Rigaku, Cu K , θ/θ diffractometer), while the microstructure and chemical composition of the electrode was examined using a scanning electron microscope (Philips XL30, equipped with EDX). The microstructure of the powder was analyzed using a transmission electron microscope (Philips EM 420). In addition, electron energy-loss spectroscopy (EELS) was used to map the elements at the nano-scale using a high-resolution transmission electron microscope (HRTEM, JEOL 4000EX).

RESULT AND DISCUSSIOIN

Preliminary experiments conducted on Si/TiN composites comprising a molar ratio of Si:TiN = 1:2 (i.e. 33.3 mol% Si) exhibited the best electrochemical properties. Hence, the rest of the work focused on this particular composition. In order to analyze the phases present after milling, x-ray diffraction was conducted on the as-milled powders obtained after milling for various time periods (see Fig 1(a)). All the peaks in the patterns correspond to TiN and the broad nature of the peaks are clearly indicative of the nanocrystalline nature of the nitride. An estimate of the crystallite sizes was made by measuring the width of the XRD peaks and using them in the Scherer equation.[15] Results indicate that the crystallites are in the 5 ~ 7 nm range. Peak broadening due to residual internal strain was not accounted for, and hence the actual size of the crystallites can be expected to be larger. The non-observance of any silicon related peak in XRD indicates that silicon exists in an x-ray amorphous form well dispersed inside the powder even after milling for only 6 h. This suggests that the HEMM process provides enough energy to generate the nanocomposite powder of silicon and TiN, although it is not clear whether silicon reacts with TiN to form any Si-N type bonds. More extensive structural characterization using MASS-NMR may be necessary to identify the existence of these environments. Nevertheless, based on the XRD patterns, it can be convincingly argued that the composites are composed of nanosized TiN containing a uniform dispersion of silicon independent of the composition. The very small size of the silicon crystals prevents the identification of a distinct phase boundary between silicon and TiN. As a result, it is possible that there is a continuous change in volume rather than abrupt discrete transitions normally observed. Thus its influence on the neighboring inactive phase is minimized. This is in fact, an important requirement for achieving good cyclability.

The specific gravimetric and the equivalent volumetric capacity of the electrode prepared with these powders are shown in Fig. 1 (b). The gravimetric capacity was converted to the equivalent volumetric form using the calculated density of the Si/TiN composite. The overall capacity appears to decrease as the milling time is increased, indicating a reduction in the amount of the active silicon phase. The exact reason for the decrease in capacity is still not clear and further detailed characterization studies are needed. The reasons at present could be speculated. One possible reason could be that the silicon nanoparticles are embedded or enclosed by TiN during milling, thereby preventing their reaction with lithium. The composite obtained after milling for 6 h shows fade in capacity, while the samples milled for a longer time exhibit moderate to good capacity retention. H. Li, et al. have investigated Si/carbon black composite for potential use as an anode.[16] The material although exhibits an initial capacity as high as ≈ 3000 mAh/g, however shows poor retention characteristics. This could be attributed to the poor binding between the active and inactive components.

The composite obtained after milling for 12 h exhibits an initial discharge capacity of 422mAh/g but a good stable average capacity of 300 mAh/g (see Fig. 1 (b)). The overall capacity is still lower than the theoretical capacity of 776 mAh/g, calculated assuming complete reaction of silicon with 4.4 Li. Although this data corresponds to the reaction of only 1.7 Li atoms per single atom of silicon, this does not necessarily suggest the formation of Li_xSi (x=1.7) due to a large (≈ 44%) fraction of inactive silicon as mentioned above. The composite exhibits a lower gravimetric capacity in comparison to conventional carbon, however, it exhibits a ~30% higher volumetric capacity, reflecting its promising nature.

Fig. 2 shows the variation of the cell potential with time for all the twenty cycles for the composite containing 33.3 mol% silicon obtained after milling for 12 h. The plot indicates that this anode composition exhibits a smooth plateau in the low voltage range without exhibiting significant fade in capacity. The difference on the time axis between the first and subsequent cycles suggests the first irreversible capacity loss (~30%). The exact reason for the irreversible loss is still unclear at the present stage. More detailed studies would be necessary to provide further insight into this problem. Some of the more obvious reasons could be the formation of Li-containing passivation layer and/or possible oxidation of the surface of the composite. Studies to assess these reasons are currently on-going.

An SEM micrograph and the energy dispersive x-ray (EDX) map of the 12h milled powder containing 33.3 mol% silicon are shown in Fig. 3. The particles are agglomerated although they are extremely small in the range of 100~500 nm. All these particles are therefore true composites containing an intimate mixture of silicon and TiN according to the EDX analysis. Thus, there is no evidence of any crystalline Si peaks in the XRD patterns. The EDX results also show the presence of iron (≈3.3%) in the as-milled powder, arising from the vial or the balls used during milling. The relatively small amount of iron can be assumed to have no detrimental influence on the capacity.

In order to analyze any changes in the microstructure or morphology of the particles during cycling, the particles before and after cycling were observed under the SEM. Fig. 4 (a) shows the morphologies of the electrodes fabricated from the composites containing 33.3 mol% of silicon obtained after milling for 12h. The surface of the electrode after 30 cycles is devoid of any cracks, which are typically observed in other metal-based alloys that are used as anodes. Moreover, there appears to be no change in the morphology of the particles before and after cycling, which indicates

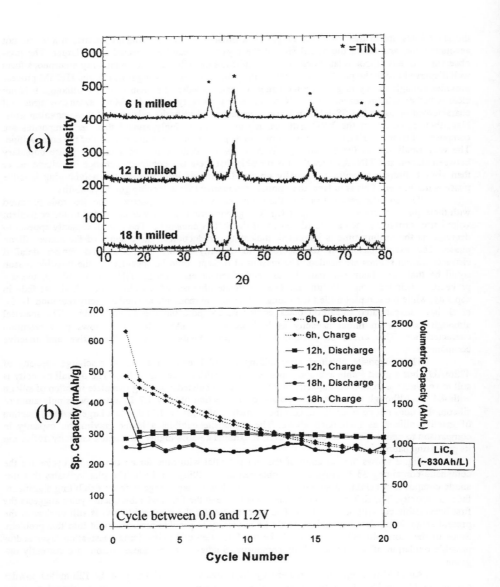

Figure 1. (a) X-ray diffraction patterns of Si:TiN = 1:2 composites milled for 6 h, 12 h and 18 h, respectively. (b) Capacity as a function of cycle number for Si/TiN nanocomposites obtained after milling for 6 h, 12 h and 18 h each.

Materials for Electrochemical Energy Conversion and Storage

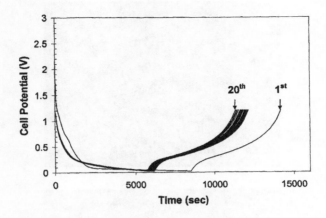

Figure 2. Cell potential profile for the first twenty cycles of the 12 h milled Si/TiN composite with Si:TiN = 1:2 molar ratio obtained after milling for 12h.

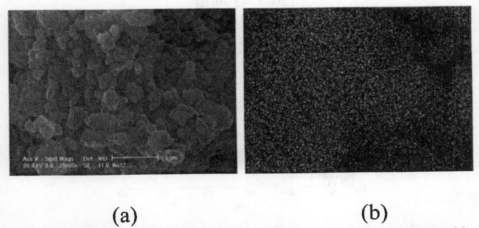

(a)   (b)

Figure 3. (a) SEM micrograph of the Si:TiN = 1:2 composite showing the nanocrystalline particles. (b) Chemical map of Si using EDX for the Si/TiN composite with Si:TiN = 1:2 molar ratio obtained after milling for 12 h. (Both images are taken at the same scale.)

After cycling

(a)

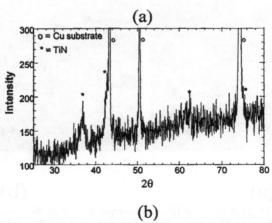

(b)

Figure 4. (a) SEM micrographs of the electrode before and after 30 cycles, (b) XRD pattern of the electrode after 30 cycles prepared with the Si/TiN composite corresponding to Si:TiN = 1:2 molar ratio obtained after milling for 12 h.

the good stability of the composite. The XRD pattern obtained on the electrode after testing for 30 cycles presented in Fig. 4 (b) is identical to Fig. 1. This suggests that silicon remains as very fine amorphous particles existing without undergoing any crystallographic phase change. The excellent stability of the electrode may be attributed to the existence of very finely dispersed amorphous silicon particles within the nanocrystalline TiN matrix. These results suggest that TiN indeed appears to be a good inactive matrix for Li-ion anodes. Furthermore, the results imply the potential of Si/TiN nanocomposites as anodes for Li-ion applications.

Transmission electron microscopy was conducted on this powder to analyze the nanocrystalline state of the composite. Figs. 5 (a) and (c) show the TEM image and selected area diffraction pattern of the 12 h milled powder containing 33.3 mol% silicon, respectively. The bright field image indicates the presence of very fine size crystallites (< 20 nm) within a single particle, indicating the true nanocomposite nature of this material. The well-defined rings in the selected area diffraction pattern correspond to the nanocrystallites of TiN with no trace of silicon being observed. This suggests that the silicon contained in the composite is amorphous as indicated by the XRD result. Fig. 5 (b) is the dark field image corresponding to the crystalline TiN phase. Comparing the region marked with an arrow in the DF and the BF images, it appears that the dark spots in the BF image correspond to TiN crystallites. Although these TEM results suggest that the composite be exactly comprised of nanosized crystallites of TiN and amorphous silicon, it is difficult to ascertain the distribution of both phases. Therefore, high-resolution transmission electron microscopy was conducted on the composite using EELS to map the distribution of the elements at the nanoscale.

Fig. 6 (a) shows the high-resolution image of the composite containing 33 mol% silicon obtained after milling for 12 h. The dark areas represent the TiN nano-crystallites while the bright regions correspond to amorphous silicon as mentioned above. Lattice fringes from the TiN particles can be observed conforming to the nanocrystalline nature of TiN. Figs. 6 (b) and (c) show the high-resolution elemental maps corresponding to silicon and Ti, respectively obtained using electron energy-loss spectroscopy (EELS). The bright regions in the map indicate the high intensity or content of each component. Elemental maps show that TiN is distributed homogeneously in the powder and silicon appears to surround the TiN particles. This suggests the influence of mechanical milling on the two materials exhibiting different hardnesses. Silicon exhibits a lower hardness in comparison to TiN and thus undergoes significant pulverization. Thus the TiN particles appear to be uniformly coated with finely milled amorphous silicon forming a nanocomposite. Generally, the samples need to be sufficiently thin for obtaining good electron energy-loss spectra.[17] However, due to the difficulties associated with sample preparation, the as-milled powder was directly used in this case. As a result, overlapping of the maps of each component can be expected and the boundaries are also possibly diffused. Nevertheless, the elemental maps do reveal two different regions separated distinctly to represent the distribution of silicon and TiN phases. These results are therefore indicative of the stability of the Si/TiN nanocomposite arising from the nanoscale distribution of the two phases achieved by mechanical milling. Results of these studies therefore reveal two aspects, the promise of Si/TiN nanocomposite as a useful electrochemical system and the potential of mechanical milling for synthesizing electrochemically active nanocomposite materials.

CONCLUSIONS

Nanostructured Si/TiN composite powders produced by HEMM are comprised of amorphous silicon and nanosized TiN. As the amount of active silicon is reduced or the milling time is increased, a reduction in the initial specific capacity was observed. The exact reason for reduction in activity is not known. The milling time appears to control the amount of active silicon exposed to lithium, and prolonged milling appears to cause an increase in the inactive portion of silicon. The electrode containing 33.3 mol% silicon, milled for 12 h shows good capacity (~300 mAh/g or ~1100 Ah/L) with little fade (~0.36% /cycle). The as-milled powder consists of agglomerates of nanosized particles, each single particle itself representing a nanocomposite of silicon and TiN as indicated by the EDX results. The electrode structure is also very stable during cycling because no cracking/crumbling and/or obvious clustering of silicon was observed after 30 cycles. The nanocomposite is composed of TiN nanocrystallites and amorphous silicon according to TEM analysis. Nanoscale elemental mapping

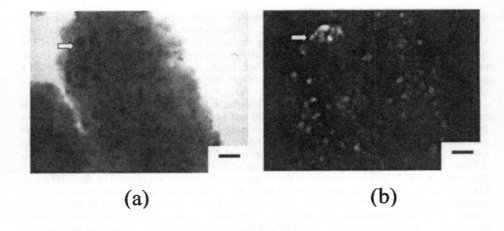

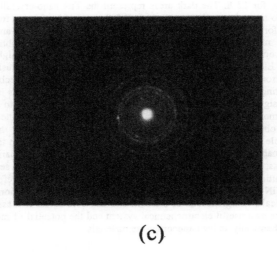

Figure 5. TEM micrographs of the Si/TiN nanocomposite containing 33 mol% Si obtained after milling for 12 h; (a) BF image, (b) DF image, and (c) SA diffraction pattern (camera length = 66cm, reduced to 30% of its original size).

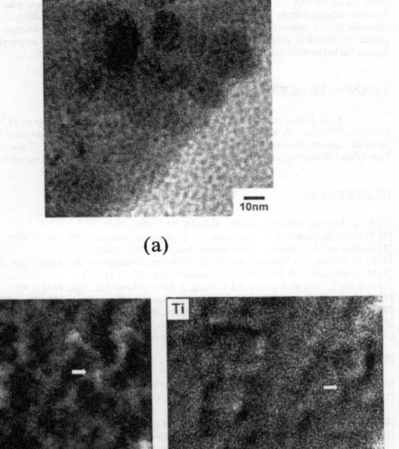

Figure 6. (a) HRTEM micrographs of the Si/TiN nanocomposite containing 33 mol% Si obtained after milling for 12 h, (b) Elemental map of Si, (c) Elemental map of Ti, which are analyzed by electron energy-loss spectroscopy (EELS). (The arrows represent the same position.)

using EELS reveals the uniform and homogeneous distribution of the two phases. These results therefore suggest that the Si/TiN nanocomposite certainly appears to be promising as an anode material although further optimization studies need to be conducted in order to demonstrate its optimum properties. Detailed structural and electrochemical studies are currently in progress and will be reported in subsequent publications.

ACKNOWLEDGEMENTS

P. N. Kumta and Il-seok Kim would like to acknowledge the support of NSF (CTS Grants 9700343, 0000563). P. N. Kumta, Il-seok Kim and G. E. Blomgren would also like to thank the financial support of ONR (Grant N00014-00-1-0516). Changs Ascending (Taiwan) and Pittsburgh Plate Glass (Pittsburgh) are also acknowledged for providing partial financial support.

REFERENCES

[1] R. A. Huggins, *Solid State Ionics*, **113-115**, 57 (1998).
[2] Y. Idota, A. Matsufuji, Y. Maekawa, and T. Miyasaki, *Science*, **276**, 1395 (1997).
[3] O. Mao and J. R. Dahn, *J. Electrochem. Soc.*, **146**, 423 (1999).
[4] K. D. Kepler, J. T. Vaughey, and M. M. Thakeray, *Electrochem. Solid-State Lett.*, **2**, 307 (1999).
[5] M. Winter and J. O. Besenhard, *Electrochim. Acta*, **45**, 31 (1999).
[6] H. Kim, J. Choi, H. J. Sohn, and T. Kang, *J. Electrochem. Soc.*, **146**, 4401 (1999).
[7] H. Zheng, K. Oka, and J. D. Mackenzie, *Mat. Res. Soc. Symp. Proc.*, **271** 893 (1992).
[8] T. Granziani and A. Bellosi, *J. Mater. Sci. Lett.*, **14**, 1078 (1995).
[9] K. Kamiya and T. Nishijima, *J. Am. Ceram. Soc.*, **73**, 2750 (1990).
[10] C. F. Windisch, Jr., J. W. Virden, S. H. Elder, J. Liu, and M. H. Engelhard, *J. Electrochem. Soc.*, **145**, 1211 (1998).
[11] M. R. Wixom, D. J. Tarnowski, J. M. Parker, J. Q. Lee, P. L. Chen, I. Song, and L. T. Thompson, *Mat. Res. Soc. Symp. Proc.*, **496**, 643 (1997).
[12] I. S. Kim and P. N. Kumta, unpublished data (1999).
[13] E. Gaffet, F. Bernard. J. C. Niepce, F. Charlot, C. Gras, G. L. Caer, J. L. Guichard, P. Delcroix, A. Mocellin, and O. Tillement, *J. Mater. Chem.*, **9**, 305 (1999).
[14] I. S. Kim and P. N. Kumta, *Electrochem. Solid-State Lett.*, **3**, 493 (2000).
[15] B. D. Cullity, *Elements of X-ray Diffraction, Second Ed.*, p. 284, Addison-Wesley, MA (1978).
[16] H. Li, X. Huang, L. Chen Z. Wu, and Y. Liang, *Electrochem. Solid-State Lett.*, **2**, 547 (1999).
[17] D. B. Williams and C. B. Carter, *Transmission Electron Microscopy*, p. 655, Plenum Press, NY (1996)

KEYWORD AND AUTHOR INDEX

Materials for Electrochemical Energy Conversion and Storage

Printed and bound by CPI Group (UK) Ltd, Croydon, CR0 4YY

16/04/2025

14658452-0001